Disclaimer

The Professional and Higher Partnership Ltd has no responsibility for the persistence or accuracy of URLs for external or third-party websites referred to in this publication, and does not guarantee that any content on such websites is, or will remain, accurate or appropriate.

The material contained in this publication is provided in good faith as general guidance. The advice and strategies contained herein may not be suitable for every situation. No liability can be accepted by The Professional and Higher Partnership Ltd for any liability, loss, risk, or damage which is incurred as a consequence, whether direct or indirect, of using or applying any of the contents of this book or the advice or guidance contained therein.

The publisher and the author make no warranties or representations with respect to the completeness or accuracy of the contents of this work and specifically disclaim all warranties, including without limitation warranties of fitness for a particular purpose. No warranty may be created or extended by sales or promotional materials.

53

interesting ways to
communicate
your research

Edited by Irenee Daly
and Aoife Brophy Haney

P&H

ISBN: 978-1-907076-64-0 (ePub edition)
       978-1-907076-65-7 (PDF edition)
       978-1-907076-63-3 (paperback edition)

Published under The Professional and Higher Partnership imprint
by The Professional and Higher Partnership Ltd
Registered office: Suite 7, Lyndon House, 8 King's Court,
Willie Snaith Road, Newmarket, Suffolk, CB8 7SG, UK

Company website: http://pandhp.com

This edition published 2014.

Credits
Abstract: Anthony Haynes
Copy-editing: Karen Haynes
Cover design: Benn Linfield (www.bennlinfield.com)
Cover image: Rika Newcombe (www.rikanewcombe.co.uk)
Proofreading: Richard Kitchen

Contents

Abstract

To be an effective researcher one needs both to conduct high quality research and to communicate it. Research may be communicated to a variety of stakeholders including specialists, researchers in other fields, business, government, the third sector, and the public. A range of methods is available, including presentation, publication, and new and traditional forms of media. 53 practical ideas, rooted in experience, are provided. Overall, the text is designed to help reflective practitioners in professional, scholarly, or scientific research prosper.

Key terms: communication; impact; presenting; publication; public engagement; research; social media; writing.

Professional and Higher Education: series information

Titles in the Professional and Higher Education series include:

Publishers' foreword

Until now, our Professional and Higher Education series has focused entirely on teaching and learning. *53 interesting ways to communicate your research* signals our decision to extend the series to cover other aspects of work in post-compulsory education.

While broadening the series, our intention is to preserve its original values. Each book in the series provides constructive ideas that are rooted in practice and readily applicable. Overall this book is designed as a supportive resource for researchers working in professional, academic, or scientific settings.

Anthony Haynes and Karen Haynes
The Professional and Higher Partnership

Editors' preface

As researchers, we often focus so much on the methods and rigour of our research that we forget to think about how we're going to communicate with the right audiences. And even when we do give our audiences some thought, we may leave it to the end of a project to really get started rather than build in a communication strategy from the beginning.

This gives rise to two main problems. First, we miss out on valuable opportunities to get feedback on our work at various stages and to join in the right conversations in our fields. Research, with its ivory tower image, is often viewed as a lonely activity. But in truth some of the best research ideas come from communicating with others. One need only reflect on the satisfaction of chatting with a like-minded colleague at a conference or at a departmental tea break to appreciate this. Communicating and sharing new knowledge is as much at the heart of research as is its discovery.

Second, while researchers have always had to get their contribution to knowledge 'out there', the days of doing this solely via conferences and journal articles are disappearing. It is no longer enough to reach a small audience who read certain journals or attend certain conferences. Research councils and other grant agencies around the world are looking for more and more value from their research funds. As a result, research is now being assessed in terms of its impact beyond academic publications. This means engaging with different kinds of audiences. In this new environment, learning how to communicate ideas to a multitude of audiences is an important part of developing as a researcher. These audiences range from co-authors and people at conferences to policymakers, members of the general public and even potential employers.

The good news is that there has never been a time when it has been so easy to reach out to so many, and at such low cost. But doing so in a way that helps you as a researcher requires some thought. As this book will demonstrate, communication is not just something to think about when you've got that paper fully drafted. By taking away the emphasis on the 'final product', the opportunities to communicate research can begin much sooner in your research career. They can also help to clarify the importance of your work for society and may even lead to involving audiences in your research activities.

We have put together a selection of 53 practical and imaginative ways for you to communicate your research, supplied, tried and tested by a fantastic group of authors (see Notes on contributors at the end of this book). Some of the ideas are focused on doing the traditional things well and in new ways, i.e. the things that are a researcher's bread and butter like journal or book publications and presenting at conferences. For example, there are tips on how to make your poster stand out from the crowd and how to make a presentation more dynamic through the judicious use of images and objects.

Other ideas are focused on old and new means of engaging with academic and non-academic audiences, for example using social media and traditional media outlets to expand your reach. Learn how to position your research for a newspaper commentary piece or a radio show. Find out how to devise a social media strategy for yourself and prepare to broaden your public engagement horizons with stand-up comedy, drama and more. In the age of the internet getting your research noticed goes hand in hand with building your own personal profile as a researcher. This book helps you to navigate the world of LinkedIn and Academia.edu, as well as to present yourself more effectively to employers in CVs and cover letters.

53 interesting ways to communicate your research is designed for dipping into, not for reading from start to finish. You can create your own path through the book in a number of ways. To start with, we have divided the book into three main chapters: 1. Communicating within academia; 2. Communicating beyond academia; and 3. General techniques. The contents page lists the 53 ideas according to these sections. We have also compiled a thematic index that

links similar ideas. Here you can find, for example, all the contributions that deal with careers or with the visual side of presenting. Finally, at the end of each contribution, you will find a list of suggested complementary ideas. These are ideas that build on what you have just read. So for instance, if you have just finished reading about peer review, you will find a suggestion to consider looking at 'Sharing your research process via social media' and 'Writing engaging lay summaries'.

Just as you will find your own path through the book, we expect that you will also implement the ideas in your own way. With this in mind, we hope that the print and e-book editions of these 53 ideas will act as a springboard for many more ideas to follow. Our intention is to implement the ideas collected here and to continue to communicate them with you, our audience, in an online format. To hear more from us and the other contributors and most importantly to share your journey, please follow us on Twitter (@researchology) and think about telling us how you've implemented a new way of communicating your research through our website: www.researchology.org

Irenee Daly and Aoife Brophy Haney

About the editors

Dr Irenee Daly is a psychology lecturer at DeMontfort University where she is also a member of the Reproduction Research Group. Her research focuses on how women decide to have children in the context of reproductive technologies and later first-time motherhood. She received her BA in 2001 from Trinity College Dublin, and an MSc in 2003 from the University of Durham. The beginning of her PhD marked an interest in learning to communicate in an engaging and clear fashion. However, watching the rapid rise of social media and its increasing influence on academia made her realise the way that we communicate to our research audience has changed. Until now she has dabbled at the fringes, but is currently putting into practice all she has learnt during this project. In 2012 Irenee received her PhD from the Centre for Family Research at the University of Cambridge. She still lives in Cambridge with her husband Donal, their Jack Russell Barney, and the newest family member, baby Quinn.

Dr Aoife Brophy Haney is a researcher, writer and academic. She is a senior researcher with SusTec, the Chair of Sustainability and Technology, at ETH University in Zürich. Aoife received her BA from Trinity College Dublin in 2003, her MA from the School of Advanced International Studies, Johns Hopkins University in 2007 and completed her PhD at the Judge Business School, University of Cambridge in 2013. She has always been drawn to the craft of research: the combination of knowledge with the creative process of developing research questions, choosing appropriate methods and telling the story of the results. Aoife experiments with new ways of improving her own research process and facilitates workshops for postgraduate students on how to think differently about research. Since the beginning of her academic career, she has been convinced that better communication improves the quality of research as well as its impact on society. Born and bred in Dublin, Aoife is an accomplished clarinet and piano player, and now lives in Zürich with her husband Ryan.

Thematic index

Chapter 1

Communicating within academia

1 Posters – a graphical research connection

2 Abstracts – more than a final thought

3 Academic interviews

4 Key terms – make them work for you

5 Twitter as a conferencing tool

6 Conference networking – building a net that works

7 Responding to peer reviews

8 Turn your thesis into a book

9 Copyright – know where you stand

10 Webinars

11 Share a conference paper using YouTube

12 Share your research process via social media

13 Open access publishing

14 How to use your research in a lecture

15 Turn your thesis into something other than a book

16 How to teach project management using your research

1 Posters – a graphical research connection

Of all methods of research dissemination, posters are probably the least understood. So, before you start copying and pasting, realise that posters are about networking, feedback and sharing ideas as a way of starting conversations. They're *not* purely about data, prose and references. That's what published papers are for. So if you want an audience to engage, attract them and make them curious.

Start by reducing your core message into *one sentence*. The average length of time that a reader spends on a poster is reported to be about two minutes. That's an awfully short time to read the 'entire thesis on one sheet' you'll see regularly at conferences. Rather than try to whittle your work down, build up the draft content from one point.

Next, ask how graphical your research is. Can you *show* your work (in which case your total limit is about 400 words) or is it essentially discursive and text-based? If so, be careful. The poster is *still* essentially a graphical medium; limit yourself to no more than 750 words.

Once you've considered your purpose and disciplinary constraints, now think about the audience. As an audience member, I want you to do four key things.

Captivate me
Get my attention; both graphically and intellectually. Why is your specialty interesting and why is it vital *now*? Captivate me and I'll read for longer. (Also, remember that I'll make a commitment decision from some distance. Do your efforts look attractive? Is your title in large enough font, not too long and jargon free? Can you state it as a question?)

Tell me a simple story

Briefly show me your story. You don't need an abstract, full methods or extensive references. (Why not give a URL, or QR code to access your methodology and/or references in full?) Think about the story you are telling. In the sciences, the majority of posters are sub-titled 'Introduction, Methods, Results, Discussion, Conclusion' (and are often ignored by a non-captive audience on a lunch-break). There's nothing wrong with this essential scaffold, but it doesn't make anyone curious. What questions does your work deal with? Place these questions, as subtitles, in a logical order, ask them and then provide a concise answer.

Just show me the essentials

Don't show me every result, just the vital ones. Use colour to highlight and show key findings. I've just spent three hours concentrating on presentations – don't make me work too hard.

Help me to connect with you

Finally, consider that a poster should also be about *possibilities* instead of finite conclusions. What's the interesting future question or approach that's going to pique stimulating conversations with the audience? Take the pressure off, and hand them some possibilities to discuss with you.

Design these four elements into an attractive story that you can show concisely to someone who doesn't have your jargon shortcuts. Finally, give the audience an A4 version (with a business card attached) to take away.

If you graphically captivate them with the simple essentials of your work and ensure conversations, then maybe your connection will last for more than two minutes.

Steve Hutchinson

Complementary ideas include 4, 6, 32, 34, 39, 46, 52 and 53.

2 Abstracts – more than a final thought

Your abstract is more important than your paper. I repeat, more important. Typically, abstracts may be accessed independently of the paper itself – in conference programmes, for example, or on journal websites or databases. Often readers will read the abstract but not the paper. And decisions are made on the basis of abstracts – for example, whether to accept your conference submission or whether to access the text of your paper. Yet frequently abstracts are less well written than the actual papers.

Here's how to write a good abstract. First, identify the various types of component that you might include in your abstract. They include:

1) the context of your research (either the scholarly context or the practical);
2) the problem or challenge that motivated your study;
3) the research question;
4) the aim of the study;
5) the methodology;
6) your results;
7) the practical implications of your study;
8) the implications for theory or for further research.

An abstract does not necessarily need to include all of these components. In particular, (2), (3), and (4) might overlap with each other: often two out of these three will prove sufficient. (5) and (6), however, are usually essential. If you do decide to omit any of the above, I recommend not changing the relative order of those that remain: for example, it's better not to place (5) before (4). Preserving the relative order makes it easier to ensure that your abstract tells a story.

Write your abstract *before* you write your paper – then, as you draft the text of the paper, keep returning to the abstract to review and revise it. Use crowdsourcing to help you edit the abstract before submitting it: asks lots of people to read it (abstracts are so short, they can hardly refuse). Give them limited tasks – for example, 'Please improve one word for me'.

Above all, remember that the abstract is *not* part of your paper. The abstract needs to be self-standing. Think of it not as an introduction to your paper but as a crystallisation of it.

Anthony Haynes

Complementary ideas include 4, 21 and 49.

3 Academic interviews

The single most important way to communicate your research during an academic job interview is to focus on why the work needs to be done. This is an old adage, but it merits serious consideration. It contains a simple lesson which is all too often overlooked or misinterpreted. Employers are looking for research that is more than simply interesting, more than simply novel or different; they are looking to support work for which there is a clear and present requirement. In other words, success in pitching your research during job interviews comes down to *need*. Before getting into the detail of your idea, its technical intricacies and your reasons why you are the best researcher to carry it out, you have to convince employers that your project is not a 'nice to have' but a 'must have'.

The traditional definition of a PhD – that it should be an original contribution to knowledge – is still very much current, but this definition can be unhelpful when pitching your research in an interview. Put simply, this is because originality is not the same thing as necessity. Many researchers are prone to confusing these two categories when describing their work in interviews: 'Nobody has researched this topic before; I am researching this topic now; therefore my research must be important and necessary.' Unfortunately, this argument is pure sophistry. Similarly misleading is the much-overused expression that the research is 'the first book-length study of [insert topic here]'. In both cases, selection panels will likely have the same question: why does it matter that this gap exists? This question can, and will, be restated in any number of ways. What is the nature of the problem presented by the lack of substantial work on this topic? What have we been getting wrong until now? What have we hitherto been unable to do? What will we be able to do differently once your research has been completed?

Nowadays, the hot topic of *impact* – whether economic, societal, or cultural – can cause a wealth of anxieties in researchers, who fear that the only way to argue for the value of their research is to pitch its impact on a vast scale – to suggest, in other words, that the life of the average person on the street will be forever altered by the outcomes of the research. Of course, it's fabulous if these things are true and you are in a position to show that your work has the power to change the world, but, in truth, comparatively few researchers can argue this convincingly. But don't overlook the potential of your work to change what other researchers and scholars are doing, because this is a type of impact that academic interview panels will understand. Whose work will need to take account of yours? What will they need to do differently?

Practise talking about why your work *needs* to done. And mind the gap, because it is not the same thing as a need.

Steve Joy

Complementary ideas include 19, 20, 21, 22, 28, 30, 32, 37 and 41.

4 Key terms – make them work for you

At long last, you've finished writing your paper. And the abstract. Thank goodness it's done. Only, oh! – you realise you've forgotten to provide a list of the key terms (or keywords) to accompany the abstract. Well, that's not difficult: it's just a few words or phrases, right? You dash them off and submit the paper.

Whoa! Key terms are in fact very important. Crucially, they affect the discoverability of your paper: the more your terms coincide with the search terms that readers who would be interested in your research will use, the more readers will discover your paper – and in turn, access, read, and cite it.

Moreover, key terms help you to position *yourself* as a researcher. It's much easier to claim to be, say, the guy-who-uses-the-Delphi-method-to-research-scenarios-in-the-automobile-industry if you've actually included 'Delphi method', 'scenarios', and 'automobile industry' among your key terms. Your selection of key terms helps to define your personal brand – for example, by indicating whether you're a theoretician, say, or an empirical researcher. Ultimately you can use your characteristic key terms to generate your elevator pitch (encapsulating the research and its worth in less than a minute).

You can optimise your selection of key terms by working on it over a longer timescale. Build up, and keep refining, a list of candidate terms (say, one or two dozen) from the very start of your research career (or, failing that, as soon as you read this!). Asterisk the most important ones.

Then, when you turn to work on a specific paper, draft a list of the key terms for that paper before you start writing it. If the conference or journal you're writing for limits the number of terms you're allowed, maximise discoverability by using the maximum number permitted – avoid falling short of the limit. Then, while you're writing the text of the paper, keep returning to the list to review and revise it.

There's a (very natural) danger that you will list only words that are important to you, rather than the terms that are most important to your prospective readers. Empathise with those readers. What terms do they think in? What terms will resonate in their heads? This is particularly important in interdisciplinary contexts, where the key terms for the discipline(s) you're talking to may not be those of the discipline(s) you're talking from. For example, something that may be called a 'framework' in one discipline may be called, say, a 'model' or 'tool' in another.

You can educate yourself about all this by, for example:

- analysing the language used in the call for papers you're responding to;
- researching the frequency of key terms used in the journal(s) you're targeting;
- scanning indexes of books, word clouds on blogs, etc.;
- identifying the key terms in the descriptions of research positions that you wish to apply for.

In addition, I would recommend, if you have not already done so, playing with Google Alerts. This is a tool that can provide you (at regular intervals) with detailed information on the currency and usage of your selected terms. It helps you to gauge how popular each term is (at least in relative terms) and also to see what contexts each term is used in.

Anthony Haynes

Complementary ideas include 2, 6, 36 and 46.

5 Twitter as a conferencing tool

Conferences provide researchers with fantastic opportunities to meet like-minded people, to interact with leaders in their field and hear the 'breaking news'. Twitter, the microblogging social media site, can help you to squeeze far more value from conferences – even when you can't attend them.

Taking your first steps
To find the right people and organisations to connect with before and during the conference, you need a clear strategy. Ask yourself:

- What do I want to talk about?
- Which people and organisations are involved in the conference?
- Who would I like to meet face to face, but don't know?

Translation please!
Even with a clear goal, the first steps to tweeting are still mysterious, largely because of the terminology. This becomes transparent with time (there are a number of helpful blogs which demystify the terms) but to get started you'll need to recognise:

- @ – username of a person or organisation and their address for messages;
- # – the hashtag is a label which identifies a topic, event or theme;
- RT – a repeat posting (re-tweet) of someone else's message – an easy form of dissemination from one stream to another.

Before the conference
When you first sign in and enter Twitter, your stream (list of messages) will be empty. There are many ways to build your stream, but for a conference you should start by looking for the key personalities in your field who are speaking at the event. You don't need to know a person to follow them and you don't need to follow people back if they follow you (a quick scan of recent

tweets will usually reveal if this is worthwhile). Many senior academics engage with their followers on Twitter – if you mention them in a tweet of your own (particularly in positive terms!) and they acknowledge this, it could give you the confidence to approach them in person.

Conference organisers may use Twitter to publicise workshops, highlight interesting speakers and disseminate key messages from the programme. Contact them in advance to ask what hashtag (#) they are using for the event. This gives a shortcut to a stream of comments, highlights and perspectives from the event. You can tap into some of the conference 'buzz' even if you aren't there!

The hashtag is also used to label messages linked to particular issues as well as events. #phdchat, #ecrchat and #acwri address general researchers' interests and with a little investigation you'll identify those which relate to your specific interests.

Finding your voice
If your experience of using Twitter at conferences is positive, you will probably want to use it more widely. Your communication style should be authentic and will develop with time, but Twitter is commonly used to: signpost to resources (journal article, blog post, useful website); publicise an event (seminar, conference, workshop); react to something (a news article, conference presentation, current topic); ask for help. Remember that it is a public arena (even if you protect your tweets) so you should think carefully about what you choose to post. Never jeopardise future publications and understand that anything could be read by future employers or collaborators. For most people Twitter is full of supportive and like-minded people; it's a place where researchers go to engage in debate and share their enthusiasms.

Sara Shinton

Complementary ideas include 4, 6, 12, 19, 25, 33, 36, 47 and 49.

6 Conference networking – building a net that works

Seven seconds. It's often said that a huge amount of residual impression is left within seven seconds of meeting someone for the first time. So when you are at a conference and someone asks you who you are and what you do, it's probably wise to have planned your answer. After all, at an academic conference, most people will have the same basic 'I'm a (e.g.) physicist' internal response. So, do you have a distinctive and memorable introduction that can be expressed in a sentence (for when you meet the vice-chancellor in a lift), in a paragraph and in a page? Practise these to ensure that they are jargon free and allow you to share your passion for the subject. Start with 'why', which is far more likely to connect to another area than the 'what' you do.

In a joined-up, digitally-connected research community it is vital that we are well networked, but for many researchers the word 'networking' conjures up images of 'working a room', small talk and schmoozing Professor Big for a job. And it's these beliefs that can make networking hard. Yet the research world thrives on collaboration, and academic meetings are not about data – they're about people. If conferences were all about data, they could happen electronically. Everyone is there for the same reason – even the most eminent academic. Simply realising this makes a networking approach a little easier.

Firstly, you need to do some research. Who will be at the meeting? Use the conference's administrative support systems (or even the Twitter hashtag) to find out who will be in attendance. Go for quality not quantity. Who are the five people you want to talk to over the course of the conference? Then, think about who is already in your 'net'. Who do you know already? Is there anyone who could provide an introduction or help initiate a conversation?

If you struggle with the neutral opening lines and academic small talk, you may wish to craft an email in advance of the conference saying that you'd appreciate a brief conversation. Given that established academics use conference down-time to discuss work with co-authors and meet collaborative partners, sometimes the only way to make an approach is to do it in advance of the meeting.

If this seems forward, then remember that provided that you are polite, respectful, and stick to the appropriate cultural conventions, then really you have nothing to lose. Even if they reject you outright, you are in exactly the same position as you were at the start. And why would you want to meet or collaborate with someone who is dismissive of someone they don't yet know? If the thought of making an approach to an established figure still seems intimidating then consider what you have to offer them. It is very difficult to beg a favour from a stranger, but very easy to offer one. And if they say 'no', they're not rejecting you – just your offer. So what's in it for them?

Make your default setting 'How can we collaborate?' when you meet someone. Don't ask this question explicitly but be looking for synergies and shared interests with everyone you meet. You'll find you start asking more open and interested questions. And if you combine that with an interesting research summary sentence you have the ingredients for building a 'net' that actually 'works'. Then, all you have to do is act on it.

Steve Hutchinson

Complementary ideas include 5, 12, 21, 32, 33, 36, 37, 43, 47 and 52.

7 Responding to peer reviews

So you've submitted your research paper, the editor didn't reject it outright, and all you have to do is answer the reviewers' comments. This could be relatively straightforward or extremely challenging, depending on the nature of the comments. This is one of the most important forms of research communication; not only can this letter be the difference between your article being published or not, it can also affect the quality of the final manuscript. You need to ensure that, in addressing the concerns of the reviewers, you also enhance your work rather than detract from the paper's key message.

Unfortunately, it's also a form of communication where researchers get very little guidance, and rarely see each other's work. You can develop your writing by asking senior colleagues to comment on your response letters, or asking to read some of theirs, or by acting as a reviewer yourself. However, there are some general tips that can help.

When you receive the comments, it's worth reading them through, alongside the manuscript, and then waiting a while before you respond. This gives you time to consider what has been said and makes it less likely that you will be defensive. To start, paste the editor's and reviewers' comments into a document, then use this to structure your letter by responding to each of their points in turn. The editor usually gives some guidance about which of these points are the most important to consider.

In terms of your responses, you essentially have three choices. First, you may agree with the reviewer, in which case you just need to describe how you are addressing the criticism, and make a note of where in the manuscript the change has occurred. Secondly, you may make a partial concession, where you perhaps acknowledge a limitation more explicitly in the text, but don't make the full alteration suggested. Finally, you may feel, on careful reflection, that the suggestion is incorrect or would not enhance the manuscript.

Whatever decision you make, you will need to make a persuasive case to the editor to explain your actions, particularly if you are disputing a suggestion; in this case, it can be useful to ask yourself a few questions to guide your response. For example, if the reviewer has misunderstood, can you see why and could you clarify this in the manuscript? Is there precedent for your approach that you could use to defend your decisions? Does the reviewer reveal a fundamental difference in their theories or assumptions that you can use to argue that their comments are not relevant? Are there some small concessions you can make to show that you are open to the review process, to make it easier to stand your ground on more central issues? Or, finally, are the comments so at odds with the paper that you would rather stand your ground and risk rejection, to retain the integrity of your research? Whatever you conclude, take the time to articulate a clear, measured argument that avoids emotive language and makes clear that you respect the editorial process.

Once you have addressed all of the comments, do take the time to re-read your revised manuscript for consistency (e.g. if you've removed a section, have you removed all reference to it throughout?) and to ensure that your key message is still clear and coherent. You can also make very minor editorial changes at this stage, although you shouldn't do any major redrafts. The last thing to do is to preface your responses with a polite letter to the editor, thanking them for their support in reviewing the paper, and then keep your fingers crossed!

Victoria Burns

Complementary ideas include 12 and 21.

8 Turn your thesis into a book

Theses and books may look similar – typically both consist mostly of expository prose and extend to, say, 70,000 words. Yet they are in fact different species. A thesis is written for a small number of specialist readers who, as supervisors and examiners, are obliged to read it. It is intended to demonstrate that its author deserves a higher degree. A book, in contrast – even a monograph (i.e. a scholarly tome) – is written for a much larger, diverse, body of readers, none of whom has to read it, still less assess the author as a candidate for a formal qualification. Yet the similarities between a thesis and book may lead an author to overlook the differences and hence to underestimate the amount of labour typically required to transform the genre of the work.

The solution to this problem lies in making a psychological shift. It is natural for the author of a thesis to think, at the outset, 'What's the minimum I need to do to turn this into a book?' But the outcome of a gritted-teeth approach of this kind tends to satisfy neither author nor publisher (nor, if it gets that far, reader). So adopt a new mindset: forget altogether about the thesis and focus instead on the question, 'What would make a really good book?' When you've answered that question fully, then you can turn to your thesis and ask, 'How can I use this to help me?'

To develop the necessary mindset, think as concretely as possible. Ask yourself, which bookshelf – in the bookshop or the library – would your book go on? What books would it sit next to? Who would publish it and in which series? What style of cover and text design would the publisher use? What guidelines does the publisher provide for authors? Try to *see* the book in your mind's eye.

There is a danger, of course, that the thesis will keep reasserting itself, trying to muscle the book out of pride of place in your imagination. After all, you're leaving your thesis – after a long, intense, and likely monogamous relationship – for another text: it's natural for the thesis to feel jealous. One way to avoid this danger is to construct a plan comprising three lists: (1) what material from your thesis you are going to leave out of the book; (2) what material you are going to include in the book that wasn't in the thesis; and (3) what you are going to change around. Ensure that each list includes genuinely substantial items.

One strategy you may wish to consider is to generate two books from your research, one for a specialist audience and one for a general readership. Consider, for example, the approach used by Professor Alister McGrath when he researched the life and work of C. S. Lewis. He produced a biography – C. S. Lewis: a life – published by a trade press for the general reader. The book uses a minimum of jargon and eschews esoteric scholarly debate. He reserved his more specialised, scholarly material for a second book – The intellectual world of C. S. Lewis – published by an academic press as a monograph. Such a strategy may serve to maximise the communicativeness of the research.

Bibliography
McGrath, A., C. S. Lewis: a life (London: Hodder & Stoughton, 2013).
McGrath, A., The intellectual world of C. S. Lewis (Chichester: Wiley-Blackwell, 2014).

Anthony Haynes

Complementary ideas include 9, 13, 15 and 34.

9 Copyright – know where you stand

As a researcher you want to communicate your work as widely as possible, but publishing your research in a journal article or a book has implications for your ownership of the work and what you can do with it. You need to observe the copyright laws in the preparation of your own publication and consider what rights you would like to keep before you sign a publishing contract. If you would like to post a copy of your article on a social media platform such as Academia.edu or share it with your students make sure you have the right to do so.

Who owns the copyright in your work?
As an academic author, you start as the copyright owner of your work (teaching materials being a possible exception – the university might own them). Copyright laws differ between countries, but basically they give the copyright owner the exclusive right to reproduce the work, to publish and to communicate the work to the public (for example: Copyright Act 1968 (Cth), Australia; The Copyright, Designs and Patents Act 1988, UK; Copyright Act of 1976, US). When you sign a publishing contract you transfer some or all of your copyright to the publisher.

Publishing contracts vary but they usually stipulate the timelines, who handles permissions and foreign rights, how electronic and future book editions will be handled, what happens when the book goes out of print (make sure full rights revert to you). Publishers typically want to retain exclusive rights to publish, sell and sub-license the work in order to protect their commercial interests and defend authors' rights against plagiarism, copyright infringement or unauthorised use.

Can you retain your copyright?
Most book publishing contracts no longer require assignation of copyright (except for textbooks). Instead they grant an exclusive

or non-exclusive licence to publish, which is a publishing agreement between the author and the publisher. The publisher gets permission to reproduce and communicate the work, but does not transfer the author's copyright to the publisher.

Many journal publishers retain full rights or place an embargo on your work if you need to comply with an open access mandate, allowing you to distribute it after an agreed time. In contrast, open access journal publishers typically release articles under the Creative Commons licenses which allow others to 'use, distribute and create their own work from yours', provided you are credited as the author (see http://creativecommons.org/licenses/).

What are your moral rights?
While copyright is designed to protect the 'economic rights' of copyright owners, moral rights protect the reputation and integrity of creators. Treatment of moral rights differs across the world. In Australia, for example, moral rights cannot be assigned or transferred, while in the US and the UK they can be waived. You should always retain your moral rights, which basically means that you have the right to be identified as the author of the work and have the right to object to derogatory treatment.

Don't forget to obtain appropriate permissions
When you are preparing your work for a publication, note that obtaining permissions is the author's responsibility. For research and literary works, copyright lasts for seventy years after the author's death. A fair dealing provision allows you to quote limited portions of a work without permission for purposes of criticism, reviewing, teaching and research. Written permissions may need to be obtained, however, for the use of figures, tables, illustrations, graphics and poems.

Agata Mrva-Montoya

Complementary ideas include 6, 8, 13 and 15.

10 Webinars

Universities want their staff to undertake international collaborations and yet they have less money to pay for international travel. These two facts of academic life make it important to use innovative ways of disseminating research. If you participate in a scholarly group that wishes to share research but cannot afford to gather for a meeting, webinar technology may be the solution to your problems. In teaching, it can be used for seminar meetings for distance learning students or those with mobility difficulties.

A webinar is a web-seminar, a virtual meeting involving sound and image that links a large number of participants in different locations. A webinar is better than a video conference because hundreds of people could be involved in a webinar whereas video conferencing is limited to a small group.

In order to host a webinar you need to join a webinar hosting site; many are free, so do a Google search and find out if there's an appropriate package for you. An example is www.anymeeting.com.

Some packages provide audio and slides only, while others allow a live video feed to be transmitted. Free sites usually only host small numbers of participants and it will also cost extra to use software that records your webinar for later use.

As with a face-to-face seminar, a speaker will present using Power-Point slides (or other presentation software) and then take questions from the audience, although for smaller webinars a round-table moderated discussion is also possible. Each contributor will need a computer connected to broadband, a camera and a microphone. For a large webinar, having technical assistants on hand to answer questions while you are leading or presenting is invaluable. If you have contributors from across the globe, bear in mind international time differences when choosing a mutually convenient time and date for your webinar. As webinar leader you will be responsible for advertising, marketing and creating the content. Create slides summarising the aims of the session and introducing each participant before they speak and prepare a minute-by-minute agenda so that you always know who will be leading the content. Keep it short: participants' attention spans will not stretch for much more than an hour.

Webinars do have their problems. Some contributors may be reluctant to participate because they worry about not having enough technical knowledge and so they may need some reassuring. Indeed, technical issues can be a significant barrier as webcams and microphones sometimes malfunction, delaying the contribution of a participant. The other main issue is that holding a face-to-face meeting creates a dynamic that is not replicable by any other form of contact. The informal chat with other delegates that occurs during break times can help to cement relationships and you can network more easily through the privacy of a personal discussion. Having said that, webinars do have something to offer scholars who want to share their research with a global audience or who want to use innovative teaching methods.

Catherine Armstrong

Complementary ideas include 19, 32, 33, 34, 39, 40, 42, 43, 45 and 52.

11 Share a conference paper using YouTube

Academic researchers often write a paper for a conference, deliver a presentation to a number of delegates and perhaps see the paper appear afterwards in the conference publication. The process is confined to the event and its accompanying materials, which may be online but are often still paper based. Academic research projects may result in several such papers and presentations which could be used to inform and support the work of other professionals in the field – but who never hear about them.

One way to publicise and share research papers more widely is to use electronic tools and social media. A voiced-over video of a presentation is a long-lasting resource which can be accessed by many, disseminated by various means (e.g. linked to your other social media sites), and can enable comment and discussion for a considerable time. The voice-over makes the video more accessible than a standard presentation, and more clearly expresses the ideas of the presenter than slides alone. Most potential audiences are familiar with YouTube and other video sharing websites. Videos and papers can be published online under a Creative Commons licence.

To make a video, save presentation slides as a series of .jpg images. Import these into a free video editing tool like Movie Maker (PC) or iMovie (Mac) which are standard installations. Arrange the images in order to make a video and add a voice-over using the inbuilt tool or a free sound editing tool like Audacity. Alternatively, use conversion tools in your presentation software to create a video with voice-over. Upload it to YouTube (or another video sharing site) where it can be linked to other websites. Link the video to your blog or profile page, or create a webpage. Upload Word or PDF versions of the paper, so that visitors are able to access all the information after they have seen the video. Share the link to the webpage containing the video and paper using Facebook, Twitter, email and other appropriate social networking tools. Ask for comments from visitors and social contacts, and enable comments on your webpage or blog and on your YouTube channel. Peer review is an important part of publication and this is one way to gather feedback.

There are a few things to be aware of at the start. Be careful not to make the video too long. Five minutes is long enough to engage someone's attention, so only talk about the key point on each slide. Slides should not contain more than a few words, and preferably just an image. People can read the paper afterwards to get the full detail. Make a transcript of the video and upload it to YouTube if you want to make the presentation more accessible. There are free tools available like CaptionTube, and YouTube has an auto transcription service. Make sure that images or music used in the video comply with copyright law. Finally, check the webpages now and again for feedback and comments.

Esther Barrett

Complementary ideas include 10, 32, 33 36, 40 and 45.

12 Share your research process via social media

The final result of months of research almost always exceeds the amount you can share in a few thousand words through a journal paper. The knowledge about the skills that you learned along the way, the many iterations you needed to make your experiment work – these stories are limited to the fellow researchers in your institution. Others, who might be running into the same problems, could be spending months trying to solve the same issues you experienced earlier on. The main problem, in fact, is that there is no defined platform for publishing and sharing the more practical aspects of research.

Online research networks and blogs can become, and in fact are becoming, a place where researchers share the story behind the final research result. From personal narratives on selecting the right population for a questionnaire, via sharing technical drawings of test setups to pieces of programming code – the number of possible elements that can be shared online is limitless.

To reach out to like-minded peers and researchers at the same stage of their career, Twitter provides an excellent tool. You can expand your network by browsing through Twitter lists of, for example, 'PhD students' or 'Engineers' or by following tweeps from the list of an acquaintance based on their Twitter bio. Additionally, you can join a Twitter chat and become active in the online community associated with this chat (such as #phdchat, #ecrchat or #acwri) where you can learn tips and tricks from more experienced researchers. Mailing list servers work similarly, and although a bit more old-fashioned, these are a goldmine for programmers sharing code – ask your coding senior grad student friends about recommended lists to subscribe to.

After gaining some first ideas on Twitter, you can delve deeper and use blogs and online forums to describe the practical aspects of your experiments in more detail, as these platforms allow for longer contributions. You can find relevant online forums by jotting a few keywords (such as: 'Fortran programmers forum') into your search engine. (Before you sign up, check the submission dates of posts of a forum to see if it's still active.) By sharing your blog entries on Twitter, you can direct your network to your post and have a discussion in the comments section.

Upon publication of a paper, you can share it on academic networking sites such as Academia.edu, LinkedIn or ResearchGate. This provides a platform not only for sharing your papers, but also for providing supplemental resources – datasets, raw data, media files or negative results. After making this additional documentation available, you can blog and/or tweet about it so that the network you developed stays informed and is invited to keep the discussion going.

Using new technologies to move research forward is not without risks in terms of ownership and copyright. As no centralised platforms are available yet, it is up to the users to form, define and manage their online research communities. Know that there are options to protect your work and ideas. For example, if you use a blog to flesh out your ideas, then add a copyright statement. If you feel uncomfortable about sharing unpublished research, wait until your first conference paper on the topic is accepted before you share information online.

Eva Lantsoght

Complementary ideas include 5, 6, 35, 36, 38, 47, 48 and 49.

12 *Share your research process via social media*

13 Open access publishing

The open access (OA) publishing model combines the traditional process of scholarly publishing with the advantages of digital technology and online delivery to ensure access to research, which is free of charge to the reader and can be free of most copyright and licensing restrictions.

Decide if open access is right for you

If you are a recipient of a grant, you may need to publish your research in OA in order to comply with the requirements of your funding body. Even if you don't need to comply, there are many advantages in making your research available in OA including increased downloads and citations, and greater visibility. If you are hoping to earn royalties on your book, OA publishing is not for you.

If open access appeals to you, consider – in no particular order – the following steps.

Identify a journal or publisher

It is important to choose the publisher wisely – the Directory of Open Access Journals publishes a list of reputable OA publishers, and you can also check the Directory of Open Access Books. Apart from dedicated OA journal and book publishers, many commercial publishing houses offer a 'hybrid OA' model. This gives authors the option to make their work available in OA on payment of a fee (commonly called an Article Processing Charge or APC).

Secure the funding

Some OA publishers do not charge authors at all – typically those who rely on institutional subsidies. But commercial and some OA publishers require authors to pay publication fees or contribute to the cost of managing peer review, copy-editing, typesetting, and hosting the content of OA publications. These fees can be high and you may need to find a source of funding at your university or elsewhere.

13 *Open access publishing*

Many research funding organisations, especially those that mandate OA publishing, allow for grant funds to be used to pay publication fees (you need to take this into account when preparing a budget in grant applications). If you cannot secure the required fees, you can often apply for these to be reduced or waived. You can also try crowdsourcing.

Choose between the gold and green OA models

In the gold OA model, the publisher makes your work freely available upon publication. In the green OA model the researcher publishes in the conventional manner (not incurring a publication fee) and deposits a copy of the peer-reviewed work in an OA repository. The green OA method therefore relies on the publisher giving permission to deposit preprints or final copies of the work in an OA repository, which can be institutional (at your university), subject based (such as arXiv) or universal (such as Academia.edu). In green OA publishers usually stipulate an embargo period, typically lasting 12 months.

Review copyright agreements and licensing restrictions

OA publications are protected by copyright, but authors are encouraged to allow the reuse of their work by others under the Creative Commons licenses (http://creativecommons.org/licenses/) such as CC-BY (Attribution) and CC-BY-NC (Noncommercial). If you are publishing with a commercial publisher and planning to deposit your work in an OA repository or a social networking site, retain the moral right and copyright in your contract and negotiate the right to distribute your works after an embargo period.

Making your research available through open access publishing means that anyone with an internet connection can download, read and use the research, facilitating its wide dissemination, transparency and higher research productivity.

Agata Mrva-Montoya

Complementary ideas include 9, 15, 19, 36 and 49.

14 How to use your research in a lecture

I'll never forget the feeling of shock and confusion that accompanied the most terrifying lecture of my life. On that day, our human physiology lesson was meant to introduce the basic principles that underlie the electrocardiogram, a clinical test that assesses heart health. Unfortunately, the lecturer mistook his audience and instead of a basic overview, we were bombarded with a two-hour acronym-laden exposition on his latest research about electrocardiogram...something or other. Needless to say, this inappropriately expert-level instruction left all of us first year medical students completely overwhelmed and thoroughly unprepared to interpret an actual electrocardiogram.

As researchers with teaching responsibilities, it's tempting for us to stick with what we strive to know and love best – our own research findings and the cutting-edge developments in our field. While these elements undoubtedly enrich student learning, the pitfalls that plagued my nightmare electrocardiogram lecture can be avoided with the following three steps:

1. Know your audience
Be able to answer the following questions: What are the prerequisite courses for enrolled students? Which topics have already been covered in the course? Has another lecturer already taught a lesson on similar material? What are students meant to take away from your lecture? Knowing the audience helps the lecturer titrate the lesson's objectives and level of detail to maximise student learning opportunities.

2. Highlight basic principles
You love your research. Your students may not. Most students on the course do not aspire to become a specialist in your field. Instead, they hope to learn the overarching principles of the field and apply them to new situations, perhaps even outside of academia. Based on your knowledge of the audience, review the basic tenets of the material

with a focus on connections between your lecture and the rest of the field. Most importantly, emphasise the reasoning that underlies the principles, so students are prepared to tackle novel problems.

3. Teach a toolkit

With appropriate background in place, research findings can be used to illustrate the underlying fundamentals. Instead of bombarding unsuspecting students with data and details, focus on the scientific process that drives your work. How did you arrive at your question? Why is your approach well suited to study the problem? If your methods are complicated, explain them in the context of the basic principles and propose a hypothetical set of results for students to interpret. What logical steps drove your choice of follow-up experiments, interpretation of results, and ultimate conclusions? How do your findings fit within the framework of the field? Strategic use of research to highlight basic principles increases students' understanding of your work. More importantly, it underscores the key concepts and provides real-life examples of the academic process.

Brennan Decker

Complementary ideas include 16, 26, 34, 39, 42, 43, 45 and 52.

15 Turn your thesis into something other than a book

On completing a thesis, many researchers turn their thoughts to publication. One option is to convert the thesis into a monograph. In some settings, that option may be foisted upon you – publishing a book may be a requirement for gaining tenure (or even for gaining your degree). Where that isn't the case, the option may not necessarily appeal. After all, converting a thesis into a monograph typically involves a good deal of labour.

So what are the alternatives? One option is to publish the thesis unreconstructed. There are two ways to do this. You can publish it in your university's online repository. One advantage of doing so is that the university is in effect lending its brand to the work (and users may be happier to download a file from a repository rather than from, say, your personal website). In addition, as a repository item your thesis will start to show up in various databases.

The second way to publish the unreconstructed thesis is to use a commercial service. If you do this, ensure you choose a reputable one. Perhaps the best-known service is ProQuest's Dissertation Express.

Alternatively, you can mine your thesis for journal papers. If you do, be aware that a paper is a different animal from a chapter: some reworking will be required. Ensure that each paper (a) is self-contained, (b) says enough about the context, (c) includes an explicit statement of the problem (or challenge) that motivated your research and of the aim of the paper itself.

There are two pitfalls here, both concerning intellectual property:
1. In your thesis, you may have used someone else's figure (a graph, say, or a flow chart). If so, ensure you have copyright permission before publishing it. Acknowledgement is not sufficient, however full and accurate the citation. (And, yes, being included in a repository does count as publication.)

2. Journals exist primarily to advance knowledge. Typically, they don't want to publish material that's been published before — a point re-enforced by their standard author contracts. If you publish your thesis first, it may then be difficult to publish journal papers derived from it: if a paper's based at all closely on a part of your thesis, you may not be able to claim truthfully that it's new material. If your goal is to publish journal papers, the safest route is to publish them either *instead of* or *before* the thesis.

Anthony Haynes

Complementary ideas include 8, 9, 13, 21 and 34.

16 How to teach project management using your research

Teaching is an opportunity to share skills as well as knowledge. Help your students to acquire project management skills by running a session based on your research experience. Your research will benefit from this critical appraisal and your students will be better prepared to tackle their own research projects.

Run this session when students are planning their research projects and ask them to prepare research ideas in advance. You will need to examine your research successes and difficulties, honestly assessing whether projects could have been managed more effectively. Ask yourself what are the valuable research planning skills that you wish you had learnt earlier. Break the session up into activities based on the key skills you identify; a few examples follow but adapt the session to reflect the skills that are particularly relevant to your subject.

1. Define research objectives
Ask your students to brainstorm all possible objectives for their projects. Now use your research to demonstrate the importance of objectives being SMART (specific, measurable, achievable, realistic, timely). Include some objectives that did not fulfil SMART principles and describe the problems that occurred as a result. Ask your students to select and defend their objectives to the group with reference to SMART.

2. Identify research outputs
The primary output for students is a dissertation, but their work could also be submitted for prizes, conferences and publication. Present your written and oral work including examples of excellent and unsuccessful research communication. Analyse what you have learnt about selecting effective outputs and give your top tips to the group. Ask your students to identify additional outputs for their work with justification of their format and audience choices.

3. Create a timetable

Simple linear timetables can be inefficient in research as one problem can hold up the whole research process. Instead, experiment with free project management software in the session to create Gantt charts, mind maps and network diagrams for your research and for that of your students. Ask your students to formulate a full plan in their preferred format after the session.

4. Risk analysis

Failures are common and an important part of the research process. Present failures that you have experienced and analyse how you moved on from these to achieve success. Ask each student to assess what might go wrong with their project and then work together to help develop potential solutions. Creating strategies in advance will enable students to cope better with setbacks.

Following the session, ask your students to complete a research project plan and hold a one-to-one session with each student to discuss the plan and assess the feasibility of their project. You now have a template for monitoring the progress of your students' research projects.

In short, structured planning of research maximises the chance of success. Use your research experience as a tool in teaching to emphasise this to your students. The outcome will be increased focus and productivity for both you and your students.

Eleanor Carter

Complementary ideas include 14, 42, 43, 45, 52 and 53.

Chapter 2

Communicating beyond academia

17 Writing op-eds

An excellent way to publicise research is through daily newspapers, in the op-ed (i.e. opposite the editorial page) or commentary sections. Op-ed writing tends to be formulaic, and the basic flow can be grasped by reading a few. Op-eds are short, usually 800 words or fewer. They are often pegged to the news of the day. For instance, if rising oil prices are in the news, you might see an op-ed advocating renewable energy or relaxed restrictions on offshore drilling. Op-eds usually start with a simple statement of the argument they intend to confront. The author then spends most of the rest of the article arguing why that point of view is wrong. Sometimes toward the bottom, the writer offers a policy proposal. Editors like submissions that accomplish all these things, and give bonus points when an article refutes conventional wisdom.

News hook
One of the strongest attractions that can win an editor's acceptance of your op-ed article is a link to the news of the day, also known as a 'news hook.' Since news cycles are short, and since big stories change from day to day, you need to act quickly. There is a caveat to this: if you can link your op-ed to a perennial issue of interest such as, say, government policy in the Middle East or a newsworthy discovery in healthcare, then the necessity of a hard news hook is diminished. But in general, it is best to write and send an op-ed within one day of a major news event. Editors will be looking for an expert to dissect the news for readers, and if your pitch arrives in a timely fashion you will get the job. Thus, if your research looks at aspects of mergers and acquisitions, for example, keep an eye on M&A news. If you hear a big merger announcement, you just earned the chance to tell the world about your research findings – as long as you can find a way to link them to that merger.

Timing

The best time to pitch an op-ed is ahead of a holiday period, when newspaper staffers leave for vacation and editors need copy to fill their 'news holes' for the week ahead. August is prime time for op-ed placement, as is the Christmas and New Year period. Any three-day weekend or bank holiday period brings a deficit in copy.

Submitting

Guidelines at the best newspapers with the largest readership and influence tend to say that writers should submit to one paper at a time. Papers want exclusive consideration of your work for a week, before (most likely) rejecting it and allowing you to send it elsewhere. The impracticalities of this system for a newsworthy op-ed are obvious. I argue that writers can send their pieces to two papers simultaneously. If one paper accepts your piece, you then must write to the other to retract your op-ed from consideration, with a profuse apology.

The New York Times is an example of a newspaper with clear guidelines for op-ed pieces: www.nytimes.com/content/help/site/editorial/op-ed/op-ed.html

Most newspaper websites list a form or an email address for submitting op-eds and commentaries. For information on US newspapers, see: www.theopedproject.org. You could also think about submitting a blog-pitch – see for example The Guardian's guidelines for its 'Comment is free' section at: http://www.guardian.co.uk/help/2008/jun/03/1

Jim Krane

Complementary ideas include 4, 18, 19, 20, 21 and 37.

18 Radio interviews

Radio interviews are a way of speaking directly to a non-specialist audience. As with print media, the best chance of talking about your research in this way is if you can find a link that will make a radio audience want to listen to you. For example, does your topic coincide with a national event (or anniversary of an event) that will be talked about in the media? Is there a regional connection that your local radio station would like to cover? Do your findings have the potential to solve a problem (health, social, environmental...)? Is there something unusual or bizarre about your idea, research methods or results? Is your topic a controversial one that will generate debate? If you can make a connection you are far more likely to be of interest to a broadcast journalist. For topical pieces, ensure your work is visible and your name appears at the top of web searches for your speciality. Your institution press office will have contacts – make sure they know that you are keen to participate so can put your name forward.

Recording the interview

A pre-recorded piece gives you more control and you will have more time for prior discussion with the journalist interviewing you, who will often visit your class/lab/field station to record some 'atmosphere' (background noise) to use in the final package. Remember you are speaking to a non-specialist audience so practise explaining your research in three simple sentences or less: although major hiccups can be edited out, the less work required afterwards the better.

Live interviews tend to be shorter, usually over the phone, and often as a response to a news item, so you will need to be able to think on your feet. Plan two or three key pieces of information to get across – is there a significant result, critique, or call to action you want to communicate? You should speak to the presenter or producer to test the phone line before going live, so try to get an idea of the questions (at least the first one) or the angle to expect – this may be in the 30 seconds before broadcast!

What could go wrong?

Off-topic questions

Be prepared to steer your answer back to your work (the three things you want to communicate). Reply that the question is interesting, and when you were looking at [your actual topic], you found...

Misinterpretation

Don't be afraid to politely correct errors. It's useful to provide a short summary of you and your work for the journalist to use when introducing you.

Stage fright

Remember that no-one can see you on the radio, so take notes with you (on one side of paper so you don't rustle turning pages). Write down only short words or phrases so you don't sound as if you are reading, and include a few statistics to refer to if appropriate.

Technical failure

Remember that even if you cannot hear anyone, they might still be able to hear you. Be patient – either the fault will be fixed or the interview re-scheduled.

Hannah Perrin

Complementary ideas include 4, 17, 19, 20, 32, 34, 43 and 53.

19 Public engagement

Public engagement can be described as sharing the work of research and higher education with members of society, who support research through taxation, and whose lives may be affected by the results of research.

The operative word is 'sharing' – engagement involves mutual learning and researchers also learn through their engagement practices. As long as you're prepared for conversation and not just broadcast, the possibilities for effective public engagement are extensive. Engagement and communications activities benefit from you (and colleagues) thinking through why engage, what about, who is involved or could be, how to engage and when.

Perhaps you hope to inspire more people about your research, bringing them into appreciation and questioning of your work. Maybe you want to consult people, seeking their views and input. Or perhaps you plan to collaborate with new partners.

Start with the people you're aiming to reach
Who are you interested in connecting with? It is important to research your target audience – which can consist of individuals or organisations – to avoid the danger of treating them as passive recipients of your expertise. Think about their prior knowledge and experience; find ways to meet and listen to your audience.

Sample other engagement activities
A good way to choose an engagement format is to sample other good practice. If you want to connect online with people beyond your research world, look for examples of blogs and social media where other researchers have created dialogue with a wider set of people. If you'd like to plan face-to-face engagement, see if there are public events and workshops you can attend to learn more. If you want to inspire a specific group such as young people, con-

nect with schools or other relevant organisations. Look to existing projects to find out what works and find someone who works regularly with this group (e.g. a teacher) to give you guidance.

Join in with others
One possibility is to get involved with one of the science festivals or festivals of ideas, arts and knowledge springing up all over the world. Well-received formats include interdisciplinary panel discussions with audience participation, stands staffed by researchers, exhibitions, schools' events, performances and artistic collaborations. Informal live-event formats continue year-round with science cafes, comedy nights and more. If you join an established programme, it can facilitate a wider reach and reduce the marketing burden for you as an individual researcher.

Are you receiving me?
Make sure to evaluate your engagement activities. This doesn't just mean finding out if people have enjoyed themselves. Think through why you engaged with people in the first place and whether you have achieved your aims. You or a colleague can use surveys, interviews, focus groups, observation or other methods. For example, before, during and/or after your event, you could suggest a Twitter hashtag to collect people's thoughts and feedback.

Seriously, though
In today's world, your research doesn't exist in an academic bubble but can have influence on and be influenced by a range of organisations and people in society. Your first public engagement activity can be a starting point to inspire, consult and collaborate with groups beyond academia – hopefully the start of something longer-lasting as part of your career.

A good resource for practical public engagement steps is: http://www.publicengagement.ac.uk/how

Nicola Buckley

Complementary ideas include 17, 18, 20, 24, 27, 29, 42, 47, 48 and 49.

20 Writing press releases

A press release is a means of informing local and national media of a significant event; for researchers, it is a way of disseminating important or interesting findings to a broader audience.

Format of press releases

Use headed paper. If you want the information published after a particular date, write 'For release on...' at the top; otherwise give the date and 'For immediate release'. For academic news, embargoes are usually used for information from conference papers being presented on a particular day – speak to the conference office.

Write an interesting headline. In the first paragraph, say who, what, where, when and why: this is not a whodunit, get the information in there at the start. Then provide some detail: is there something unusual, ground-breaking, or particularly controversial that you want to draw attention to? Including a few third-person sound-bites ('Dr Smith said...') that can be used as pull-out quotes is useful as it means less re-writing by the journalist is required. If you're aiming for local newspapers, emphasise the regional relevance or connection: for a first attempt, local media are generally much more accessible (and edit less) than national where the competition for space is fierce. A single-column item might be only 100 words: most published articles are between 500 and 1,000 words.

After the main text, head a section 'Information for editors' and give your full name, job title, institution, and contact details, including at least email and phone number(s). If you have collaborators, users, or beneficiaries of your research, ask them if they are willing to be contacted by the press and include their contact details if they agree. You can also include a short list of photographs you have available, or weblinks for information about your institution, department or project.

Submitting a press release

Always send your release to a named person. Read the newspaper and find a reporter who writes on your field; look up the editorial staff directory; use Twitter; or call the switchboard and ask for the newsroom or editorial department (e.g. the business, arts, science desk…). If you can make contact and pitch your story directly, you will be much more likely to be asked for details. Once you're on a journalist's contact list, you're likely to be contacted again if similar stories arise and they want an expert (or contrasting) opinion.

Potential pitfalls

Waffling
Keep releases to one side of A4 – if a reporter or editor wants a longer story, they will contact you.

Not being interesting enough
This may be your life's work, but why should anyone else care? Make sure it is news – use your institution's press office for advice.

Being uncontactable
Be sure to include your contact details and to be available at short notice if contacted by the newspaper – be by the phone.

Poor timing
Don't submit a release too late to relate it to current events – ask the paper when their print deadline is and stick to it.

Hannah Perrin

Complementary ideas include 4, 17, 18, 19, 21, 28, 36 and 37.

21 Writing engaging lay summaries

Thinking like a journalist can help you write effective lay abstracts for grant applications and journals, according to a researcher from the University of Birmingham. By making the structure and style of our writing more like a newspaper, reviewers and readers see the focus of our research, and the broader implications, within a few seconds.

Most grant agencies, and some journals, now ask scientists to produce a summary of their research for a lay audience. As well as helping non-specialists to understand your work, it gives you the opportunity to explain how your research meets the broader aims of the funders or editors. The summary also demonstrates that you can engage wider audiences, and can be used to publicise the research, if successful.

However, most researchers find this challenging. Although it is clear that you should use less technical language, and assume less background knowledge, people still tend to use a standard abstract structure; they start with the background, move on to the proposed methods, the expected results, and finish with the potential implications.

In contrast, my opening paragraph illustrates why changing the order, and adopting a more journalistic approach, can be powerful. If you read nothing else, you would have got my 'take home' message from the first few lines. Hopefully, by grabbing your attention, it also encouraged you to keep reading! It also feels less like a dumbed-down version of a 'real' abstract, and is instead written specifically to communicate effectively.

So, when taking this approach, what should you consider? Crucially, the opening line should be sufficient to get across the main point of the research. The second sentence will usually then put this in a broader context, so that the implications are clear. In essence, this should 'justify' your research in terms of its theoretical, or applied, significance to the world! By now, you should have covered the 'What, Why, How, Who, Where, and When' of the research: *what* is your research; *why* is it important; *how* will you do it; *who* is involved (researchers and/or participants); *where* it is conducted; and *when* it is conducted or the benefits realised. Now, just add more nuanced detail to these questions as the lay summary progresses. For example, this might include highlighting what is novel about the techniques ('how'), and explaining why the institution is ideal for this research ('where'). Throughout, using quotations can introduce a human voice to your summary. This makes you and your research more real, and gives the opportunity to explain directly any important issues. You can also use analogies to illustrate key concepts, and bring the research to life. Overall, remember that your reviewers are often faced with a pile of grants or articles to assess in a short time. Use the techniques developed by journalists to capture their attention and ensure that they understand and remember your work!

Victoria Burns

Complementary ideas include 4, 17, 19, 20, 22, 26 and 28.

22 Non-academic interviews

'You must kill your darlings.' This oft-repeated quote reminds writers about the dangers of gratifying their personal preferences and foibles, because pleasing yourself risks displeasing your audience. Just the same can be said of discussing your research in a non-academic context: your proudest academic achievements will not necessarily be meaningful to such employers. Can you be sure that they will know what to make of your publications, grants and awards, conference presentations, invited lectures, curricula you have devised, and so on? Your task is to decide whether these darling achievements are meaningful, and if they are not, kill them.

Speak to the employer's need
Preparing for an interview should, first and foremost, be about the employer and what is important to them. In essence, recruitment is the process whereby a business or organisation meets a need, and that need always will be distinctive to that particular organisation, at that particular moment in time. You can therefore test how meaningful your research achievements are: do they help you demonstrate that you have the skills or experience required to do what your potential employer needs of you? For example, you may be justifiably proud that you have thirteen peer-reviewed papers in top-ranking journals, but it's not self-evident that a non-academic employer would understand the achievement. Is writing a key part of the role? If so, are your papers examples of an *appropriate* sort of writing? How, specifically, do these papers help the employer see you in the job they have advertised? And are they genuinely your best evidence (defined as 'most relevant to this particular employer')?

Speak in the employer's language
Communicating your relevant experience is part of what makes you eligible to do a job, but what makes you *suitable* can be harder

to pinpoint. Here, use of language is key. First, be ruthless about cutting out academic or scientific jargon. If you were asked in an interview to describe a situation in which you had solved a problem creatively, it would do you no favours to offer a highly technical, specialist explanation using vocabulary that the employer simply did not understand. Test out your language well in advance of the interview by practising with a non-expert, who can highlight any inadvertent slips into jargon.

Secondly, use the employer's own language. In other words, speak as if you were already one of the team. Planning experiments or project management? Presenting at conferences or communicating with stakeholders? Supervision or leadership? This isn't idle hair-splitting, and it isn't simply playing with buzzwords. How has the employer described their need? What vocabulary have they used? A willingness to shift your language to match the employer's will emphasise your willingness to adapt to the new environment in which you are applying to work.

Finding out about needs and language
Speaking to the employer's need and speaking in their language should not feel like a dark, intuitive art. Read their website assiduously, but, above all, speak to people who know the organisation and type of work. Use your networks and seek out opportunities to forge new contacts. Ask them to help you decide whether there are any darlings you need to kill.

Steve Joy

Complementary ideas include 3, 4, 20, 21, 30, 37, 41 and 51.

23 Information sheets – what research participants need to know

All research involving humans or organisations requires obtaining informed consent. An integral part of this process is providing the person with a Participant Information Sheet (PIS), which contains all the study details necessary to make an informed decision regarding participation. The challenge of the PIS is to capture the breadth and detail of the study in accessible language, in a relatively small amount of space.

Selling your research
Language and layout
It is vital to consider the target audience. For example, if your participants are young people the language needs to be especially straightforward; likewise for those with literacy problems or a different first language. Avoid subject-specific jargon: see http://www.plainenglish.co.uk/

Sometimes more than one PIS is required. Data collection in schools will require a PIS for the headteacher, the parents and the children, each written differently. Presenting this information in a Q&A format works well (Q: Who is carrying out this research? A: This research is being carried out by...). Sometimes this information is best conveyed as a synopsis, or even in a table or diagram.

Study aims and rationale
This is the opportunity to really 'sell' the research. This can be tricky, as the information needs to be truthful (deception as to the purpose of the study is ethically dubious), while not going into such detail as may result in the participant feeling that they have to respond/behave in a certain way during data collection (social desirability).

Part of 'selling' participation is to describe the benefits. Not all studies can or want to provide material or financial remuneration, but these are not the only types of benefit. Altruistic participation should not be underestimated, especially if the researcher can provide feedback to the participants. For example, a study about blood pressure can give valuable individual feedback about how lifestyle impacts BP. Don't raise unrealistic expectations such as suggesting that participation in your research will cure cancer. Finally, the voluntary nature of participation needs to be stressed and that it is possible to withdraw from the study both before and after the data collection.

Practicalities of participation

Where will the research take place? How long will it take and how many appointments will there be? Who needs to be present? Who is eligible for inclusion? Any inconveniences or potential discomfort however unlikely or small need to be anticipated and explained. It is wise to have a strategy for such occurrences.

Potential pitfalls

Confidentiality and anonymity

If your study promises confidentiality (nondisclosure of the person's identity) explain how you will achieve this (e.g. assigning ID numbers or reporting only grouped data).

Remember anonymity can only be guaranteed if no one (including the researchers) knows who the participants are. Online data collection methods which record IP addresses are technically not anonymous. Make sure to check the software.

Also, who will see your data? Just you? You and your supervisor? Other colleagues? You need to outline who and why. Finally it is standard to outline how the data will be used, e.g. publications and conferences, and how confidentiality will be maintained.

Data storage

Outline brief details of data storage. Specifically, is it secure? Be aware that if you store data in the cloud (e.g. Dropbox) your data is subject to the data protection laws of the country where the company have their data centre, which may be different from those in the country you are working from. Most universities have a data protection officer, who will have up-to-date information.

Irenee Daly

Complementary ideas include 12, 16 and 25.

24 Public engagement – drama meets research

How do I engage people with my research?

Communicating research effectively to the public, in an engaging way, can be a difficult prospect. It is important to make your research more relatable by stepping outside of the immediate topic area and addressing the wider impact of your research on the world. This will allow your audience to connect with the research on a personal level.

The delivery of the research message is also important. A lecture-style format, for example, is likely to engage only a limited proportion of the audience. This is particularly likely if the potential audience is young, or 'on the move', such as people walking around a museum or down a street.

Is there a solution?

Use real-world examples to bring home the relevance of your research, and do so using more interactive methods of communication. Performance art provides one such method.

To use performance art you could develop collaborations with artists across faculties in a university setting or form links with local drama clubs.

An example

Pick several real-world examples from your research topic. In the case of technology research, good examples are mummification, space exploration and global warming. Next, create a character for each of your examples, in these cases a mummy, an astronaut and a polar bear, and allow them to explain your research within the context of their lives. This personalises the application of your research and allows audiences, particularly younger ones, to engage with a character and the material that they are discussing. These characters can also be linked in an engaging manner which will challenge the audience, for example a treasure hunt format around a museum or in the street.

For this example of communicating technology research, the character of the polar bear can be used to explain the effect of modern technology on the environment through global warming. The polar bear would explain the negative impact of technology on its life but also how new 'greener' technologies are being developed to improve it.

Things to bear in mind
The collaboration
Both the artist and the researcher must ensure that the other's expectations and requirements are met to foster a healthy relationship. For example, creating a character may come second to the details of the message for the researcher, but this will allow the artist to produce an engaging and believable performance.

The audience and location
Finally, the location and nature of the audience also need to be taken into consideration. A structured and engaged audience, such as one that is seated if this activity is performed as a play, will be able to listen to a full-length script. A more transient audience, such as one walking around a museum or gallery, will require the artist to adjust the script, highlighting the most important and interesting details. This requires flexibility on the part of the artist and careful thought on the part of the writer to produce an adjustable script.

Amelia Markey

Complementary ideas include 19, 27, 34 and 42.

25 Recruiting research participants through social media

Subject areas at the interface of theory and society often face problems in looking outside the academy to find human participants for research projects. Social media platforms provide unusual opportunities to solve this problem.

Widening the participant pool: types of social media

The diverse types of social media platforms include: mixed-interest public platforms such as Facebook and Twitter, forums on special-interest and support group websites such as Mumsnet, fanpages for sports teams or musicians, and comment sections on well-established blogs and Tumblr feeds. These platforms allow researchers instant access to self-identifying members of their populations of interest. They also provide a wider demographic than the typical undergraduate participant pool, as social media users are less likely to be entirely WEIRD (White, Educated, Industrialised, Rich and Democratic).

Extending the researcher's reach

There are several ways to approach populations using social media. A researcher may set up a group specifically related to her research project and share it through her existing networks. She is not likely, however, to reach much further than her own circle of acquaintances, and might not reach significant numbers of people whose circles are very different from her own.

A more direct route uses special-interest online forums. By approaching an established community, the researcher can access existing relevant discussions and/or start her own threads oriented to the research topic, which the population of interest can access directly. The researcher is not directly dependent on other people sharing links or threads as they are already accessible to the target community.

Similarly, approaching established groups on general platforms such as Facebook allows the researcher to contact people outside of her personal networks. The researcher can post pictures, video or sound attachments as well as text to attract participants. These can then be shared by others, further extending the researcher's reach, and can form part of the methodology for data collection.

Adapting your approach according to your platform

Approaching established groups requires preparation. Groups and forums in which members regularly post updates, images and links will be more likely to participate in research and to communicate the experience to others. It is also useful to become familiar with terms, abbreviations, protocols and other types of shared knowledge which groups use, in order to minimise the impact of being 'an outsider' and to avoid potentially causing offence.

The researcher should aim, using the population's own terms, to closely associate the research aims with the group's interests and explain how the research can shed new light on that interest. Furthermore, to encourage participants to share results of research organically, it must be provided in a format appropriate for that group's practices, not just through text files. By investing more time in the presentation of research, for example by creating infographics or video clips, findings may be more readily shared within and between platforms.

Social media platforms of different kinds permit researchers direct access to a wide range of special populations. By engaging with members of these groups within their own online space, researchers may effectively gather data and share findings with lay audiences in accessible, efficient and even aesthetically pleasing ways.

Rebecca Woods and Ruwayshid Alruwaili

Complementary ideas include 5, 12, 23, 33, 44, 46 and 49.

26 Turn research outputs into stakeholder tools

Academic research outputs are generally not widely accessible to non-academic audiences. One of the reasons for this is that the commonly accepted communication method is to publish in peer-reviewed journals. Although this is a crucial component of the academic world, people outside academia do not generally read these journal papers (or, indeed, have access to them). Therefore, in addition to academic publishing, it is important to find alternative ways to make academic research more widely available to those working in fields related to and interested in your research.

One way to make research more accessible is to turn it into a 'tool'. A tool is an instrument to be used for a specific purpose. In research, tools can be used to convert complex academic work into something practical, to the benefit of the wider world. A tool can take many forms – a workshop format, checklist, database, or piece of software, for example. This perhaps only sounds suitable for a select number of academic fields (such as business or engineering), but tools can be relevant in many more areas. In the field of policy-making, say, tools can aid decision-making or offer guidelines; in the medical field a tool might be a booklet or protocol setting out a stepwise approach to testing a new drug; and in product design, a tool might be a checklist that facilitates the consideration of environmental concerns in the design process.

To build a tool, the intended audience needs to be considered carefully. How can a tool support their daily work? What do they do on a day-to-day basis? How do they prefer to retrieve information and work with others? What could make their lives easier? How can a 'tool' address this particular need? By considering these questions the right form of tool can be developed.

It is important to meet the intended audience of your research early on – whether they are in local government, hospitals, businesses or charities – as involving them in the research process may help satisfy academic criteria *and* add practical relevance. These people may be experts in their fields and can help by reading, verifying, or testing research ideas and other outputs, which may help to validate the tool and can add academic rigour.

You can expose the research further by speaking at conferences and writing extracts of your research for practitioner email lists and magazines. Proactively making these connections enables you to talk to intended users and conduct 'trial rounds' of your tool. This will not only increase your evidence base but may also lead eventually to more research projects in the area.

Last but not least: beauty! The tool needs to look appealing. If you are not a designer yourself get someone else involved to help you with this.

In short, the development of a tool, if executed well, can benefit academic rigour as well as make academic research more readily available.

Nancy Bocken

Complementary ideas include 10, 19, 28, 29, 31 and 33.

27 Stand-up comedy for researchers

Researchers often have problems connecting with the public, especially groups that do not typically engage with 'university' events. They also often find it difficult to think differently about their work, in ways that would make it more accessible to other people. One solution to this is to practise communicating research in a way that breaks down barriers between audience and speaker, positively thrives on looking at things from new angles, and for which feedback is immediate and honest – this format is stand-up comedy.

There are several ways to get started with doing research-based stand-up comedy – from signing up to an open spot in a comedy club (only recommended for people who actually want to break into comedy) to contacting the nearest Bright Club (http://www.brightclub.org/), or setting up something similar (recommended for everyone). Bright Club is a movement chiefly about comedy by researchers and many UK cities now have one, run by local groups of volunteers. They vary in operation, but typically involve six acts each doing about eight minutes of stand-up comedy about their own research in front of a paying audience that turns up to be entertained by a blend of jokes and interesting information. Shows have a professional comedian compère and set everything up to make it as easy as possible for the researchers, who usually have no prior experience of doing comedy. Bright Clubs normally provide some form of training to help with writing and performance in the lead-up to a show, and the audiences are friendly and encouraging. Bright Club is akin to doing a departmental research seminar prior to an international conference – a safe, supportive environment in which to start and get useful experience for other things. Unlike a typical departmental research seminar however, it is hugely enjoyable to do.

The most obvious difference between typical academic presentations and stand-up comedy is that, for stand-up to work, you must be the centre of attention. There are no projector slides, and no computer to hide behind. Stage presence and delivery are all important, but don't try to be someone else – take aspects of your own personality and amplify them. Observe how successful comedians stand, how they talk, how they structure their material, how they use rhythm and timing, and how they hold the microphone.

So, finally, how exactly do you do stand-up comedy about your research without appearing to devalue it, oversimplify, bend the truth, provoke controversy, or die on stage? People approach this in different ways, but one way is to treat it primarily as a public engagement exercise. What is it that you want to communicate? Pick one message, break it down into three or four elements, brainstorm keywords from those elements, and then start to think of funny things to link between those. This may seem daunting, but researchers have one huge advantage over everybody else: they all have something unique to them, and do not have to work hard to think of new jokes about well-worn topics. You may think there is nothing funny about your research, but really, if you cannot manage to find eight minutes of amusing material in what you do, then you should probably consider changing your field.

Dan Ridley-Ellis

Complementary ideas include 19, 32, 39, 42, 43 and 52.

28 Marketing your research

Suppose you want to take your research beyond academic circles: perhaps you have developed an algorithm that could transform internet search; perhaps you found an effective way to prevent HIV transmission. You now want to turn your ideas into action. This could be the creation of a new venture, product or policy. So, what is the best way to communicate with diverse audiences and avoid the usual stereotyping of academic research as too complex and 'theoretical'?

From researcher to strategist

Look at yourself not only as a researcher, but also as the Chief Strategist and Marketing Manager of your project. Before you approach anyone, develop a plan of action. The internet offers thousands of strategy and business plan templates: they are decision-making tools and will help you keep on track. Many government websites offer free advice and templates: www.gov.uk; www.sba.gov; www.business.gov.au. Skim through the different models and adapt them to suit your objectives.

Developing your plan

The content of your plan should be based on these goals and the audience you have in mind. Your plan should be able to answer questions such as: What is your work about (your idea)? What is unique and important in your findings (your unique selling point)? What do you want to achieve? Who are you aiming at (both the end user/beneficiary and any intermediaries in this process, such as funders, policymakers)? What can they lead to (the opportunity)? What are the weaknesses and likely obstacles? What about the general environment (the marketplace, competitors)? How would you go about it (strategy)? What are the resources required to make this happen (investment, people, training, equipment)?

Tailoring the plan to your audience

Adapt your content and communication style in a way that your chosen audience can relate to and find relevant to their own goals and needs. Start by identifying your audience(s): are you looking for a company to finance a new venture, a new partner or distributor? Research the sector and companies, all the way down to relevant departments and influential individuals. Trade associations and financial newspapers provide a good impartial overview, whereas LinkedIn is a great way to find out more about particular individuals (and their networks) within organisations. Understand their goals, needs and previous experiences with similar projects. Tailor your message in a way that appeals to them: this will also minimise the risk of gatekeepers turning you away. Remember, the gatekeeper can be anyone in the organisation: from a secretary who fails to connect your call, to a PR person who doesn't think your project is attractive enough.

Sharing your plan

Prepare a variety of formats for your content: a one-minute elevator pitch, an executive summary and a twenty-minute presentation. Keep complex/support data as a backup. Be ready for a one-to-one informal conversation or for a group presentation, as well as for communicating your idea via email or through your own website (always a professional point of reference). The medium of delivery is only a tool in your communication. Create a powerful impression in the first sixty seconds: focus on a clear and persuasive proposition; all the rest should support it. Be memorable through compelling content and by having established rapport. Read your audience, listen and engage in dialogue. Be clear about your aims, but also open to new ideas and opportunities. Last but not least, be personable: *people* are going to decide. And their judgment is based not only on facts, but also on how they *feel* about you.

Monica Wirz

Complementary ideas include 26, 29, 33, 34, 36, 37, 43, 51 and 52.

29 Engaging with think tanks

One key difficulty for academics is to communicate their results outside academia and to influence public debate. You don't need to set your sights just on making newspaper headlines – there is a middle way between academic and popular communication. Writing short and crisp think-tank pieces allows you to test your arguments and receive valuable feedback, get contacts and gain professional experience in publishing outside the academic environment.

Think tanks can help disseminate your research findings and recommendations

Think tanks produce ideas, research and analysis to have an impact on government policy and the public realm. They often pursue an agenda or work with a topical focus to advise on policies or their implementation. Their audiences are journalists, the wider public and decision-makers in government, companies or international organisations. Their skill is to translate complex and difficult subject matter into everyday language without the space constraints of a newspaper article but also without academic jargon.

Working with think tanks helps you to focus your messages, make your work available to a wider public and have a significant impact outside your research community. It also allows you to publish ideas swiftly and receive feedback on your work. Writing for a think tank means focusing on your findings and how they translate into advice. Policymakers are interested in scientific findings, but they need to be made clear, accessible and of relevance to them.

How to publish with a think tank

In order to publish with a think tank two things are necessary. First you need to identify and establish a relationship with a think tank in your field. You can meet them at conferences and workshops or ask colleagues collaborating with them to put you in touch. Alternatively, you can consult think tank directories and contact the expert who wrote a recent piece relating to your research. For this, it is important to understand their portfolio and how you can contribute. Think tanks produce studies and brief papers but often also host events to discuss their work with practitioners or offer them a forum to develop new ideas.

Secondly, your contribution needs to match the priorities of your new audience in ministries, companies and the media. Identify a problem and the solution and explain your argument. Ideally, your piece ends with recommendations that advance the current discussion. Realising what aspects of your research and findings are of interest takes time and thought: using friends and colleagues to brainstorm ideas often helps. Depending on your subject you could outline a critical issue related to an upcoming conference about a topic in the media, or advise, for example, how technologies can be used in education or how language skills affect employment opportunities. The point is, whatever your research, you will have important knowledge that can change the world for the better. A think tank allows you to test your arguments and ideas and receive valuable feedback without losing academic credibility. While writing for a think tank you will discover your research through new eyes and improve in communicating your scientific results to make them stick and have an impact.

Hubertus Juergenliemk

Complementary ideas include 5, 6, 19, 21, 28 and 49.

Chapter 3

General techniques

30 CVs – conversation vitae?

It can be challenging to tailor a CV when employers use the same language and phrases in adverts. Seeking advice from experts might help you to find the right focus, but it is difficult to break free of traditional CV structures. When advising researchers I ask them to imagine I am the recruiter. Putting the CV to one side, I ask them to tell me what they think I need to know to be persuaded to invite them to interview.

What would you talk about?
The conversation is clearly going to be influenced by the job and the employer.

Academic
An academic recruiter will pay great attention to where you've studied, what you've published and who you've done research with previously (as their reputation will influence the employer's opinion of you). You would talk about your role in the group, emphasising your self-reliance and initiative. Presentations at conferences demonstrate that you are trusted with the group's reputation and are able to present and defend your position.

Research in other sectors
Alongside research outputs, employers in the commercial and public sectors will need to be confident that you can work well with others, given their team-based approach. They would listen for signs that you are motivated to solve the kinds of problems they work with rather than pursuing purely intellectual problem solving. A conversation would probably go beyond your research as you might have other interests demonstrating this wider view and your ability to collaborate effectively.

Other employers
Your focus now is on the transferability and relevance of your skills rather than direct experience. Deconstructing your experiences as

a researcher, you identify the skills and learning that have value, whilst drawing in other experiences and achievements. Convince the listener of your ability to adapt to new situations, using evidence of having done this before.

So far, so good – we now compare the key points of the conversation with the CV and often find areas that need more prominence or detail. Similarly, we frequently discover details that can be thinned out or 'relegated'. My final question comes from a conversation I had many years ago with the chief executive of the Association of Graduate Recruiters. I asked him to sum up what recruiters wanted and his memorable one-word answer has stayed with me. 'Oomph.'

Where is the oomph in your CV?
The formal structure of a CV is restrictive and may leave something interesting about you tucked away in an obscurely titled section on page 2. What illustrates the type of person you are and brings more substance to your story? Starting a CV with an 'achievements' section gives you the freedom to draw the highlights together.

Back to the page
After this imaginary conversation (better still, actually talk to a friend or careers adviser), note the topics and achievements you mentioned. Return to your CV and as you put these verbal highlights into written form, a genuinely tailored document should emerge.

Sara Shinton

Complementary ideas include 3, 4, 21, 28, 36, 37, 41 and 51.

31 Podcasting

Why make a podcast?

One way of getting your research findings across to a wide audience is to create a podcast. Podcasts can be downloaded by anyone, anywhere, at any time. In addition to being relatively inexpensive and easy to create (you most likely own the equipment you need on your smartphone or laptop), you could potentially create, publish and promote your podcast in as little as an afternoon. But the greatest advantage of podcasting over other forms of engagement is that you can craft and create your message exactly as you want.

Creating a podcast – the basics

A podcast is a digital media file made available on the internet which can be downloaded to a portable media player, computer, etc. All you need is a digital recorder (or dictaphone app on your smartphone/tablet) and an audio editing application of which there are a multitude to choose from (many of which are free, such as Audacity). You can then publish your podcast on an online digital media store such as iTunes for free. Take advantage of the expertise and support around you – you may belong to an academic institution which has an established public engagement/communications office who can lend you equipment or help you publish your podcast on iTunes or another university-specific media streaming website.

Making your podcast stand out

Your podcast could be a stand-alone media item or a part of a series; it could be short (a ten-minute summary of research findings) or long (an hour-long discussion of the issue and detailed explanation of findings). It is important to bear in mind what the aim of the podcast is and whom the podcast is for. For example, if you want to communicate your research findings to as wide an audience as possible, a podcast that is relatively short (between ten and twenty minutes) may be more likely to be downloaded by the general public, whereas if you intend to target professionals in your field, a richer and more in-depth podcast may be more appropriate.

The most interesting podcasts to listen to are typically diverse in terms of texture, comprising different voices, sounds and elements. You could have a friend or colleague interview you about your work or you could record a group discussion or debate. It can be helpful to listen to established podcast series and try to emulate the styles of format that you find the most engaging.

Promoting your podcast
Once your podcast is online there are various ways to promote it. You can email a link to the podcast to colleagues both within and outside your department or university; you can use social media platforms such as Facebook and Twitter, targeting specific individuals and organisations with an interest in the field. You can also post a link to the podcast or embed the podcast on your personal webpage or platforms such as Academia.edu.

Lucy Blake

Complementary ideas include 5, 6, 9, 32, 34, 35, 36, 38 and 43.

32 Vocal exercises for presenting – speak up!

Imagine you're giving a research presentation. Your data is sound, your argument solid, your presentation a marvel of audio-visual art. Your voice, on the other hand, is often anything but prepared. There can be many reasons for a weak or quiet voice, from nerves to simply not realising the voice doesn't carry, or being unused to public speaking.

Vocal warm ups can turn a shaky, breathy, hesitant voice into a voice that projects knowledge, confidence and, crucially, can be heard at the back of the room. A few minutes spent on breathing exercises and warming up the vocal cords has a huge impact on the strength and sound of your voice. It can also help to mitigate stress and anxiety around giving presentations, helping you to sound more confident.

The first part focuses on breath. Think about how you breathe – what happens when you inhale? Chances are you pull in your stomach. Instead, concentrate on breathing from your diaphragm, expanding your ribcage to take in deep breaths. If you place your hands on your stomach, fingertips touching over your navel, in-haling should cause them to move apart. Breathe in for a count of five, exhale for five, in for six, out for six, all the way up to ten. Concentrate on filling the lungs, then exhaling all the air. This same breathing technique is used by musicians and singers to produce strong, powerful sound, and it's just as useful in the lecture hall as the concert hall.

Next warm the vocal cords by humming. Hum a nursery rhyme, a favourite song, or just scales. Start softly, and get louder. Hum for a minute or two to get the vocal cords relaxed and ready to go.

Moving on to the facial muscles – tongue, lips, cheek and jaw. Get them warmed up by repeating some of the tongue twisters below. Concentrate on articulating every syllable clearly and precisely. Repeat each one three times, getting faster if you can.

> *Sample tongue twisters*
> Rubber buggy baby bumpers
> Red lorry, yellow lorry
> She sells sea shells on the sea shore, the shells that she sells are sea shells, I'm sure.

Finish off the whole warm up with a yawn to relax the muscles. The whole exercise should take no longer than five minutes.

If you feel that you don't have time for the entire warm up there are a couple of solutions. Add five minutes onto your preparation time for the presentation (it helps to get there early anyway) and find somewhere quiet — the bathroom, a spare classroom — and go through the exercises. The breathing exercises can be practised anywhere, and are a useful habit to get into. If you only have time to practise one part of the warm up that is the most important one as it strengthens and protects the voice and helps you sound more confident.

Erika Hawkes

Complementary ideas include 11, 31, 39, 43, 44, 52 and 53.

33 Social media strategy

Whether you're using traditional ways of communicating your research (articles, books or talks), or online methods (blogs, podcasts or websites), your presence and your work may become 'lost' in the vast amount of information that's available to your intended audiences.

Strategic use of social media platforms can raise your visibility among your peers and other audiences. Social media can make you 'findable' if someone is searching for you or for a researcher 'like you' in your field, and thus help build your network. You can also keep your network of contacts and also potential contacts updated about your work and activities, and keep yourself prominently 'on their radar'. Publishers increasingly expect you to use social media to promote your work.

Explore the various social media platforms available and choose a mixture, depending on their functionality and their terms and conditions, and where your potential networks and audiences congregate. You might choose professional networks like Yammer or LinkedIn to connect with potential employers, academic social networks like ResearchGate, Methodspace or Academia.edu to share your work with peers or potential collaborators, content-sharing platforms like YouTube, Audioboo, Scribd or Slide-Share to share resources and Twitter for short, regular updates on your activities online and offline.

First, type keywords like your name or research field into a search engine and see how findable you are to valuable potential contacts, such as journal editors, collaborators, journalists or potential employers. How might people search for you, or stumble across your work? Is there anything out of date, inappropriate or unexpected? Set up a Google Alert for your name to monitor content about you added by other people.

Second, set up as full a profile as possible on your chosen social platforms, with your real name and a professional picture and 'strapline' or slogan, used consistently across your online presence. Think carefully about the keywords you use to describe yourself, as good metadata will help people search for you and form a positive, accurate impression of you.

Third, wherever possible, link your platforms, including your university personal webpage. This allows your audience to easily collate all the facets of your online professional profile, improves your search engine ranking, and by linking, for example, your Twitter, SlideShare or blog feed with other platforms, updates other sites automatically.

Fourth, don't just 'broadcast' information; interact with your network. Move from creating a static profile to connecting with institution and discipline colleagues, and then add connections of connections. Post regular small updates about your latest activities, publications and conference attendances, articles you've read recently, or responses to relevant news stories. Engage with your network: ask and answer questions, comment on posts, retweet, participate in online discussions or share resources, so it's a genuine, mutually useful exchange rather than just self-promotion.

Finally, think about how to keep your professional profile separate from your personal one, with careful use of privacy settings, and perhaps separate pseudonyms, accounts or platforms for private life. Always post with caution, even in 'private' circles, as information can become public and damage your reputation.

Helen Webster

Complementary ideas include 5, 6, 11, 12, 25, 31, 35, 36, 38, 47, 48, 49 and 50.

34 Storytelling – present your research in three acts

The trick to delivering a memorable and engaging presentation, seminar or even a public engagement lecture is not to convert your full paper(s) into slides. In fact, don't think in terms of papers at all. This is difficult to do because as researchers we become attached to the details and the structure of our papers. But these details and structure can often make it difficult for audience members to see the wood for the trees. Instead, step back and think about the story of your research.

Think in threes

Good stories have a beginning that sets the stage, a middle that piques interest and an end that resolves any tensions. Use this as a structure for your presentation. What are the three main parts of your research story? In general, the audience will want to understand what motivated the research, what problem or question you are trying to answer, and what your solution or results might be. These three elements form the backbone to many presentations.

You can divide each of these into three further points. Not only is this helpful for the audience, it gives a balance and rhythm to the presentation that will help you to remember your main points. So for instance, before you start to prepare any slides, create a matrix of the main points you want to cover. Start with three main sections and divide these into as many groups of threes as necessary. For a fifteen-minute presentation, for example, this is likely to be a 3x3 matrix: three main sections with three main points in each. This will allow you to focus more on the story and on the links between each of the three acts, rather than on the slides.

Make it your story

Instead of guiding the audience page by page through your paper, try to bring a personal touch to the story. This is your chance to explain how the paper emerged and evolved. Maybe you had a different plan originally. Or perhaps there are some anecdotes from the news, the field or from your experiments that you can use to illustrate the motivation. These may not appear in the paper but they will help your audience to connect with you and with your material. They will also make your presentation more memorable, for you and for the audience. Stories stick in people's minds.

Things to avoid

Try to avoid the standard slide at the beginning of presentations that outlines the various sections that you will discuss. Explain the structure of your story in words instead. Time is precious and presenters often waste time going through a list of items that have little meaning for the audience at the beginning of a talk.

Don't feel the need to restrict yourself to one slide for each point. Sometimes you may need more than one slide, and sometimes none at all. Indeed, there may be three main sections but you may spend significantly more time discussing the problem or the solution, depending on what you're presenting.

Aoife Brophy Haney

Complementary ideas include 1, 11, 14, 16, 18, 19, 21, 29 and 31.

35 Blogging

Books and journals are the standard platforms for communicating your research. Yet traditional publishing is slow, reaches only a limited readership and offers an outlet for only a small part of your work: your final research findings. There are many aspects of your research and life as a researcher which may be interesting to far broader audiences, from colleagues in or beyond your field to your professional peers, students or the general public. These could include:

- research 'offcuts' from publications;
- updates on a project or your activities;
- discussion of methodologies or approaches;
- commentary on related issues in the news;
- applications of your research;
- reflections and advice on being a researcher;
- reviews of conferences, books or articles;
- an online 'CV' portfolio.

Plan carefully what might best be saved for traditional publishing platforms and what might work to your advantage as a blog post.

Blogs offer an online space to communicate any aspect of your work, in the form of regularly updated 'posts' or more static 'pages'. Blogging can enhance your traditional publishing, allowing you to explore, get feedback on and stake a claim to early ideas, and promote your publications. It can also contribute to building a research community, public engagement and peer mentoring.

You can use one of many free blogging platforms to create and host your blog. Decide on the focus, purpose and audience of your blog; a well-defined remit works best (you might have several blogs for different purposes). Reading academic blogs yourself will help you decide what you might offer, and what works well. You should also decide how regularly you will update, and for how long the blog might last; if it's a project, it may have a natural lifespan. If maintaining a blog yourself seems too onerous, you could contribute guest posts to someone else's blog, or join or set up a group blog with multiple authors. Think about your publicity strategy. Adding keyword 'tags' to your posts will help search engines find them; linking your blog to your university webpage and your email signature will make people aware of it; and connecting it to other social networking platforms such as Facebook, Twitter or LinkedIn will alert people to updates.

Blog posts and journal articles are written in such a different style that in terms of copyright they can't really be said to be the same piece of work, but if you want to publish something you've blogged about previously, check with your potential publisher. You may find that maintaining a blog becomes time-consuming. If so, remember that a blog post is a totally different form to a journal article or monograph chapter; it is short (500–1000 words), deals with one small idea, and is written in accessible style without references (though you can add weblinks). If you're not gaining as many readers or comments as you'd hoped, you may need to rethink your approach to the content and presentation of your blog posts. Blogging is about dialogue, an engaging, non-academic style and exchanging useful content, not just 'broadcasting' information. Commenting on others' blogs may help to build a community of readers.

Helen Webster

Complementary ideas include 12, 19, 25, 28, 37, 38, 47, 48 and 50.

36 Online networking

In today's global research environment, there is increasing pressure to be active online. LinkedIn and Academia.edu are professional networking platforms which enable you to connect and communicate with others around the world.

So what's in it for you to engage? LinkedIn and Academia.edu can help you disseminate your research, enhance your visibility and extend your professional network, all of which are key to success as a researcher.

Adding your research papers to both platforms will enhance your Google search results, which leads to increased citations. On Academia.edu, you can look for potential collaborators among those who follow your work and the researchers you follow, while usage data can help you showcase the geographical reach of your research. By following other researchers you can also keep on top of the literature in your field.

How many times have you Googled someone you want to know more about? Having profiles on LinkedIn and Academia.edu will push your visibility higher up the Google search rankings because Google rates both of these networks highly. This will ensure people find your professional achievements (rather than your Facebook page) when they search for you.

'It's not what you know, it's who you know' is an old adage but is still true. Researchers are increasingly finding employment both within and outside academia through their professional networks. Being active on these networks will also increase your chances of finding collaborators, seeking feedback from peers, and identifying relevant funding and employment opportunities.

So how can you get the best from these networks? Here are tips for enhancing your digital identity and widening your network:

- Use the same profile photograph in all your online profiles to aid consistency and recognition.
- Complete your online profiles as they will rank higher in Google than incomplete ones.
- Keep a record of who is following your work and what keywords they are using to find you. This will demonstrate impact on your CV and inform your dissemination strategy.
- Add samples of your work to your profiles, e.g. presentations, blog posts.
- Join or create groups to expand your network and knowledge.
- Look at the profiles of others to research possible career trajectories.

Where there are benefits, there are always pitfalls and these networks are no different. There are two main limitations:

1. Time – it takes a long time to set up full profiles, keep your content current and nurture a productive network. Start small by adding the minimum for each section then set aside five minutes a week per platform to gradually add more content and connections.
2. Intellectual Property Rights (IPR) – sharing ideas and unpublished work on a public platform can lead to IPR issues. Make sure you are aware of your university's and funder's policies in this area and remember, everything online is public and permanent. You may not be able to share your results immediately, but try having a more general discourse on the topic to inform and engage others in your research.

It's worth mentioning that Academia.edu may be more useful for your day-to-day connections with other researchers. Ignoring LinkedIn is a big risk though. At the very least, use it to showcase your CV online (recruiters within and outside academia will look for you here). Preferably also use it to connect with non-academics and engage relevant groups. This will expand your knowledge beyond your current research niche and provide opportunities in areas you had never before considered.

Emma Gillaspy

Complementary ideas include 4, 5, 12, 21, 33, 37, 38, 49 and 51.

37 Professional personal profiles

In the world of 'the internet in our pocket' potential collaborators, employers, students and reviewers can all find our online presence in less time than it takes to telephone our institutions. It is vitally important therefore that we stay in control of our professional online and published presence. We know this, yet when asked to write a 'profile' for conferences, meetings, press releases, employer websites, directory websites, or simply for LinkedIn, many of us struggle to transcribe the importance of our research and the relevance of our experience into the limited word count available to us. Our profile is much more widely viewed than our CV, and we need to make sure that it is an accurate reflection of who we are and what we want to be known for.

As researchers we operate daily in a highly competitive global marketplace. By mastering our professional profile we are actively managing our global professional presence. It very quickly describes our research, our experience and our aspirations to a potentially massive audience, and as such is a hugely effective tool for researchers for networking, collaborating and career management.

So how do we fit our career, our passion and our aspirations into 200 words or fewer? By focusing on what is important. The five key questions to answer are:

1) What do you do, for whom, and what is your specific area of expertise?
2) Why is the work that you do important to the world, to your field of research and to you? Describe your emotional attachment to your research and the impact of your research on the global society
3) What qualifications, awards and accolades do you have?
4) How did you get to this point in your career? If you have worked with, or been inspired by, any of the 'giants' in your field then mention it.

5) Who are you as a human being, when not occupied with your work?

Write one sentence on each question, keeping your answers succinct and jargon free (your reader is likely to be an intelligent non-specialist).

Hone and edit your text over a few days until you have an impactful profile that is an accurate reflection of you at your best. Then read it out loud, preferably to an audience. Any part of your profile that makes you cringe or giggle needs to be changed. Remember that your profile may be read by potential employers, collaborators, colleagues, peers and people who know you well. It will undoubtedly be published, and may be available on the web for many years to come.

Your profile is important, and needs to be an accurate reflection of you, your career and your work. You need to be able to be proud of it.

Caron King

Complementary ideas include 3, 4, 6, 22, 28, 30, 33, 36, 41 and 51.

38 Multi-author blogs

Setting up an academic blog as a way of communicating research is certainly a great idea. However, running a personal blog is challenging because of the time-costs involved. Maintaining a site and enhancing its visibility are two important facets that require a significant time commitment. Setting up a multi-author blog or contributing content to a multi-author blog avoids the time-cost problem. There is a trade-off here: the former allows a higher degree of control over the process and content than the latter, but also involves more time and effort. We'll take these in turn.

How to set up a multi-author blog
Establishing a multi-author blog requires consideration of four main points. First, what is the remit of the blog? Deciding what subjects and topics – how narrow or how broad – is essential to focusing efforts and creating an identity.

Second, who is the intended audience? Is it the general public or academic peers? The former means you want to have articles that are shorter and jargon-free – or in general, accessible.

Third, once the above, important points are worked out, you have to consider various other details and auxiliary activities:

- What administering platform (i.e. what kind of software) will be used?
- What does the 'house style' look like? British or American English, etc.?
- How will the site and individual articles be publicised? It is essential to use social media platforms. Create, for example: a Twitter handle and build followers; a Facebook page; a newsletter; and so on.

Finally, as a multi-author blog necessarily entails more than one person, who and how many people will have editorial, managerial and administrative responsibilities? Here you must consider how the publishing process will work. Will each contributing author be able to self-publish or will all articles have to go through a focal point? Again there is trade-off between ease and control that has no clear answer. The latter option would make it easier to get a consistent style and voice to blog articles, but entails having a dedicated person responsible, which means hiring someone to perform the necessary tasks. This would require resources, which then brings up the question: from where?

Contributing to a multi-author blog

Instead of starting up a multi-author blog from scratch, one can become a contributor to an existing academic blogging site, of which there are many. Again, there is a loss of control over the process and over some of the finer details discussed with regard to setting up a blog. But contributing to a multi-author blog has the advantage of reducing the time-costs entailed. It also precludes having to find resources to have a dedicated staff, or to gather other people willing to set up a multi-author blog with you.

Finding a good blog to contribute to is not difficult. You want to find ones that are most closely aligned to your academic field of interest, but also should consider more 'universalist' blogs that cover a wider range of topics. This is because those are more likely to have a bigger audience and therefore reach for your research. For example, the LSE Politics and Policy blogs are narrower in focus (http://blogs.lse.ac.uk/politicsandpolicy/) than The Conversation, which is a new project that publishes academic articles on a wide variety of subject matters (http://theconversation.com/uk).

Joel Suss

Complementary ideas include 5, 12, 19, 21, 33, 35, 36, 47 and 48.

39 Presenting – know your audience

The purpose of presentations in general is to communicate information and ideas. Knowing your audience is an important first step in the process of creating an effective presentation. Your audience could comprise academics, stakeholders who have a vested interest in your work (e.g. practitioners, policymakers and professionals), as well as members of the general public. Depending on which audience you are presenting to, it is essential to tailor the content, style and delivery of your presentation accordingly. Engaging with your audience is more important than telling them everything you know in a short space of time.

The academics

Many of the academics in your audience will likely be working on a similar topic or in a related field, so they will be looking to compare your work with theirs. Nevertheless, it is always useful to briefly discuss the background and rationale for your work to set the context. In regard to content, this audience will mainly be interested in the method that you used in your research, the findings from any analyses and importantly, how they contribute to theory. Visuals in the form of clearly displayed facts and figures and graphical representations work particularly well with this group, which can help to limit your number of text slides (think quality, not quantity). The delivery of your presentation should be formal, to the point, and jargon free (if technical terms are essential, make sure that you define them). There is no hard-and-fast rule about adding 'personal touches' (e.g. anecdotes and humour) to your presentation, but make sure that they are relevant and keep it professional.

Audiences outside academia

The practitioners, policymakers and professionals
These individuals will be drawn to your presentation based on a common interest in the particular issue or problem that you are

presenting. They will primarily be interested in the 'fit' of your findings with strategic directions within their organisations, and the associated costs and benefits (i.e. the 'bottom line'). Therefore, to make the most impact, spend time reflecting on the implications of your work and ensure that you include a slide on 'recommendations for policy and practice'. In terms of style and delivery, similar to an academic audience, a 'keep-it-simple' approach is most effective here (i.e. formal and to the point using charted facts/figures and graphs). Bear in mind, these people are hoping to put your genius to work in the so-called 'real world'.

The general public
This type of audience is generally interested in the local relevance of your work. Your best strategy for impact is to be personable; use relevant examples, personal anecdotes and interesting quotations. Here, the use of images, objects, handouts and interactive activities, when possible, may be more effective than a formal presentation of statistical data. If you are presenting at a public forum or local discussion group, present in a casual (but professional) manner, using colloquial language. This is usually the most amenable type of audience so think creatively and have fun with it!

Karen Souza

Complementary ideas include 10, 11, 14, 16, 19, 34, 42, 43, 44, 45, 46, 52 and 53.

40 Images in presentations

Images are often used when sharing research and can be valuable tools. However, if used inappropriately they are an unhelpful distraction for your audience.

Choosing and using images carefully
Metaphoric images can communicate complex phenomena, assuming your target audience will understand the reference. They can be especially helpful in illustrating complex points in limited timeslots. When you choose images, remember to deal with any copyright issues. Only use an image if it is fully relevant, and it helps to express or clarify your point. Test this by removing the text you have on a slide, leaving the image. Find somebody who knows little about your work and talk through your point, then see if they make the connection with your chosen image. If they do, this may even be a good way to present your research, as text-heavy slides can be unapproachable.

Cliffhangers via VCD
While images can be self-evidently connected to the point being made, you may want to mix it up, as one catastrophe in presenting research is losing the audience's interest. Text-heavy presentations and obvious images give things away easily, but images can also generate attentiveness if used in conjunction with Visual Cognitive Dissonance (VCD). VCD occurs when the audience are perplexed by your choice of imagery. It requires your input, and therefore begs the undivided attention of your audience while you explain how the image elucidates your point.

Rule of thirds
Taken from photography, the rule of thirds helps in visually engaging an audience. Dividing your photograph (in this case, your presentation slide) up into thirds, both horizontally and vertically, may help you to think more about placement of your image. The

rule states that placement of the focal point (in this case, your image) off-centre, relative to the entire slide, will result in it being noticed more readily, while something placed in the centre is visually tedious. As well as grabbing your audience's attention, this also leaves room for any required text to be laid out in the remaining white space.

Manipulating images using computer software
Software packages provide helpful tools for those who struggle with images. Most software packages (e.g. PowerPoint, Keynote) allow you to edit images in various ways. For example you may wish to fade your image to make it less invasive, play around with the colours to make it fit in better with the rest of your presentation, or even remove a busy background that you feel is distracting from the intended focal point of your image. Others (e.g. Prezi) enable you to present ideas using a white-board style canvas that displays all your ideas in one place (that you can zoom in and out of), rather than on slides. Such a feature is especially helpful if using a detailed image to illustrate a complex point.

Two top tips
1. All images need a resolution high enough not to become pixelated when projected onto a larger screen.
2. Find a balance between use of images and use of text: variety is key for maintaining engagement.

Debbie Braybrook

Complementary ideas include 9, 10, 11, 32, 34, 42, 43, 45 and 46.

41 Cover letters that do the job

Jobs are hard to find. As a researcher you are a specialist, but you are still operating in a crowded market place. How can you increase your chances of landing the perfect job and the next step on that all-important career ladder, whether as a researcher inside academia or outside academia, or on to wider horizons?

The process is clear. You need to be invited to interview. That requires your potential employer to read your CV. And that's the job of a really good cover letter. Your cover letter is the first point of screening of your application; many recruiters will never read past the first paragraph of your letter before adding you to the reject pile. A little structure, a lot of preparation and you can make sure that your cover letter elicits the 'yes' response you need.

The cover letter serves three purposes. It tells an employer who you are and what you do, it tells them why they should care, and intrigues them enough to read your CV. In less than a page! It's not a summary of your CV and it's not a generic document. You have to write specifically for this recruiter. Keep three things in mind.

First, be very clear on what you offer. As a researcher you are a specialist, and you need to explain your knowledge in just one or two sentences. Think impact, rather than methodologies, and think tweet length. Explain why your work is important to you and to them, clearly yet succinctly. Treat them as an intelligent non-specialist and cut out the jargon. Show how your skills, knowledge and experience are relevant and evidence achievements and personal attributes to make you stand out from other researchers. Use peers and mentors to help you find these points of difference.

Second, put yourself in their shoes. It's about their needs not your potential! If you have a non-academic application tell your recruiter why and how your work is relevant to them, and for academia, why you are more suitable than the other brilliant researchers. This can be hard; taking time to think of how your biggest advocate would 'sell you' can help. Academic applications require focus on knowledge and track record, while non-academic ones need to focus on skills and relevant experience. Show them you understand what they're looking for, and respect their time by mapping the relevance of your evidence for them. At this stage point them directly to the relevant sections in your CV.

Third, ask for what you need – an interview! Tell them the post excites you and that you want to add value to them.

In terms of the basics, be clear, concise and articulate, accurate with your contact information, and address your letter to a named individual (do your research!). State why you are writing (in response to an advert, because they are doing research in your field), and be clear on why you would be great in the post.

Caron King

Complementary ideas include 3, 4, 21, 22, 28, 30, 37 and 51.

42 Objects – making your presentations memorable

The use of slides is a great homogeniser. The projector for our presentation will be in the same position as it was for the previous speaker. And so will the screen. The software too will be familiar to your audience. And so on – slides make all presenters seem alike. All of which makes it difficult to give a memorable presentation.

A way to overcome this problem is to go 3D. Instead of slides, bring some objects into the room. Doing so can offer a number of benefits. Objects arouse curiosity: seeing some fishing rods behind you will make the audience wonder, 'What are those fishing rods doing there?' And they will start trying to predict how your presentation will unfold ('How will those rods come into it? Something to do with using natural resources?') These predictions give you something to engage with.

Consider how you could use some tin cans, for example, or (to give a ragbag of examples), some bottles, a cactus, a carburettor, a clock, a grapefruit, hats, a hockey stick, rocks, a pair of spectacles, rope, or a tripod.

Objects tend to carry an aura. I've sometimes given seminars about school or college textbooks. Now, scholars who research textbooks tend to think of the books in terms of the pristine copies that arrive on their desk straight from Amazon – so I like to bring in a used textbook, with a dog-eared cover or damaged spine, and students' underlinings or graffiti. Such copies evoke something of the world they have been used in. This is particularly true of tools or utensils.

And objects are good at connecting people. They help to fill the void between speaker and audience. Speakers will often move forward into the room (towards the audience) to display their exhibits more clearly. Members of the audience come up afterwards, wanting to examine (often, feel) the object. You brought in an astrolabe: the guy in the back row, who happens to have an unrivalled collection of nineteenth-century astrolabes, is now your best friend.

You can use objects literally or metaphorically. You can use ladders or buckets to represent either nothing but themselves or, say, (respectively) levels of attainment or categories for placing ideas in.

You can make your objects talk to each other. By placing a watermelon next to a brick you can create a dialogue between the organic and the manufactured, curves and straight lines, or growth and stasis. Bring in a third object – a cup, say – and you can begin to tell a complete story (with a beginning, middle and end) about the relationships between them.

There are potential difficulties. Your object will need to be transported to the conference or lecture room. But then objects need not be large to be powerful (the bow will do, without the double bass). Thought needs to be given to health and safety, but in most cases risk assessment will be a matter of common sense. Don't load the pistol.

Anthony Haynes

Complementary ideas include 10, 14, 16, 19, 34, 43, 44, 52 and 53.

43 Taking the stress out of presenting

What is the biggest obstacle in delivering a good speech? Many times, it isn't the content. Success on stage often means being able to re-cast a seemingly insurmountable task in which you feel as if it's you alone on stage into a shared experience with your audience. Widening the stage to include your audience helps alleviate pressure from the speech itself as the sole definition of success. Combining the skill of speaking with the art of public listening can also be a useful tool to deepen your impact long after you leave the stage. And the shifts you need to make in order to present with less stress and more impact are small and manageable.

Shift 1: Acceptance – it's okay to be nervous
The first expectation that can hinder us is that we may feel we shouldn't be nervous – but it's natural to be nervous. A fearful situation, like the threat of a looming stage, will automatically invoke our body's sympathetic neurobiological response: we freeze; breathing feels difficult; our heart rate pounds. While this may be a hardwired neural reaction to a 'threat', we can change the perception that we're alone on stage by focusing on off-stage elements that can work in our favour, from the audience supporting us to how to use the time behind the curtains effectively.

Shift 2: Understand a live audience changes with you
The response of the audience can't be controlled, but their response can organically shift many times during your time on stage – as such, an audience can be very forgiving and flexible. Speeches before a live audience are never a finished narrative – they are simultaneous learning experiences for both sides of the stage.

Shift 3: Warm up
Begin informally to build your confidence and connection with the audience.

Arrive early. Take time between sessions to understand what you couldn't prepare for beforehand: by meeting your audience, learning who they are and what questions they have.

Take advantage of opportunities to connect with your audience as you take the stage – but before you begin with the 'official' part of the speech. A simple act that works well is to ask the audience what they're interested in. This has the effect of sharing the stage, helping the audience be heard, and helping you understand quickly which content may be most valuable to the audience.

Shift 4: Improvise – include your audience in your talk
Your presentation is also influenced by what your audience experienced before it. Audiences respond not just to your presentation in isolation; it's often part of a larger experience of the learning, questions and events of the day. Help the audience transition into your presentation by building upon insights from a previous talk that had the room buzzing, or 'share the stage' with the audience by acknowledging audience questions in advance and how you will address them in your talk. While audience members may want to discover something new, they also strive to connect with presentations by identifying a sense of themselves or their thinking in the work being presented – taking a minute to include this personalised context builds a useful foundation for more connection with your work.

Han Pham

Complementary ideas include 10, 11, 16, 18, 27, 32, 34, 39, 45, 52 and 53.

44 Lights, camera, conference! Using video to communicate your research

What is the one thing that you want your audience to remember from your talk? Conveying your research in a fifteen-minute conference presentation can be difficult. Video can be used to capture your audience's imagination, transforming your conference presentation from flat PowerPoint slides into a highly engaging presentation. This chapter presents four key areas to consider when planning the use of video.

Planning: key points and equipment
Before creating a video you need to be clear about your key point. This should be clearly illustrated in the video. Next, plan how to incorporate the video into the presentation and whether to record sound. Storyboard before recording, so that you know what to record and for how long. Finally, consider the technology available. Whether you have a digital camera that can capture video or a state-of-the-art video camera, familiarise yourself with the equipment and any quality settings before recording. Consider using radio mics to record conversations clearly. The higher the quality of the recording, the larger the file size and so storage may also need to be considered.

Ready to record
Recording real-world settings will not only capture your audience's interest but help them understand how to apply a solution. For example, recording a teacher's use of mobile technology with students in a classroom can be used to highlight the initial problem, show how the solution is implemented and illustrate key findings. Simulations or models may be required when a real-world setting is unavailable or prohibitive, for example to demonstrate the impact of solar flare activity.

During recording, it is important to mount any recording devices securely – you don't want make your audience motion sick! Screen-capture tools can be used to record simulations as well as computer-based research, for example distributed collaboration in virtual worlds or visualisation of data. Time-lapse photography can also be useful to record experiments or natural events.

Editing

Following recording, the next stage is to edit. Some free and easy-to-use editing software includes Windows Movie Maker, Apple iMovie and Avidemux for Linux. When editing, keep in mind your key point and critically reflect on whether this is evident in the video. Use text and transitions sparingly and consider whether sound is required, as you may arrive at the conference to discover no audio speakers are available.

Preparing to present

Finally, prepare what you will say and when you will speak: before, over or after the video. A combination of all three is often best, but beware potential pitfalls:

- Avoid describing the video in detail beforehand. Instead, highlight important points regarding the context of the recording or things to watch out for.
- When speaking over the video take care not to overload your audience. Do not provide additional information or introduce new arguments beyond what is demonstrated. Instead, highlight important points, such as aspects of the video that may be ambiguous or involve a step-by-step process.
- After the video, avoid the urge to describe what everyone has seen. Instead, draw out important points and explain how these relate to your key point.

All aspects of creating a video take time, so make sure to take this into consideration when planning. However, the impact it has on your audience can last beyond the coffee break and your research may become the talk of the conference.

Carina Girvan

Complementary ideas include 10, 11, 14, 16, 19, 26, 32 and 45.

45 Slides – rehearse the transitions

When presenters use slides, the transition between slides can be very jerky. Often there can be a clumsy pause before the presenter resumes the narrative of the presentation. The result is a lack of fluency, which can be annoying for members of the audience and risks losing their attention. The sense of hiatus is sometimes unwittingly heightened by the presenter resuming the narrative with a rather awkward, meaningless phrase such as 'OK...' (or, worse, 'So the next one's about...' – as if the presenter is being dictated to by the slides). Overall, poorly managed transitions can both bore, or even, annoy the audience and create the impression that the presentation is being driven more by the slides than by the presenter.

The solution is to rehearse the transitions between the slides. Many presenters rehearse what they are going to say about each slide but forget to rehearse the links. This, surely, is the wrong way round: if anything, the links need more rehearsal than the commentary on the content of the slides since you're less likely to know already what you're going to say in between slides. So my recommendation is: rehearse the transitions.

As you rehearse the links, focus on the big picture: use the links to emphasise the overall narrative. To maximise the fluency of the narrative, practise clicking to a new slide not at the end of a sentence, but during the course of a sentence. For example, instead of saying, 'I'd now like to explain the methodology [CLICK],' you can say 'Some aspects of the methodology [CLICK] require clarification.' Or instead of '[CLICK] OK, so, the motivation', say 'I ought first to explain why [CLICK] it was we did this project.' Clicking in the middle of a sentence, rather than at the beginning or the end, is the easiest way I know of sounding like a professional rather than an amateur. It also helps to ensure that your narrative, rather than Microsoft PowerPoint (or whichever software you use), remains centre stage.

You may find, when rehearsing the links in this way, that some of the words on the slides become redundant. This is particularly true of titles. For example, if you present your slide on methodology as you turn to the topic of methodology in your narrative (and as you use the word 'methodology'), do you really need a title on the slide saying 'Methodology'? If you experience this problem of redundancy, remove the unnecessary words from the slides. In the above example, this means removing the title – which sounds radical, but usually proves effective in practice.

Anthony Haynes

Complementary ideas include 32, 34, 39, 43 and 52.

46 Infographics – worth more than a thousand words?

To be effective, an infographic should encapsulate the full story of your research – the findings, the data, *and* the wider context. Unlike traditional data presentation where you are advised to explain in the text what your graph or table 'says', a memorable infographic that makes an impact should not require additional explanation.

What makes an infographic different and do I really need one?
Infographics are where design and data merge, but even the best ones, are tools – no more, no less. The data and your audience should always, *always*, lead the way and determine whether an infographic is required.

The key differentiators between infographics and other forms of visual presentation are context, emotion and metaphor. It can look great, but only when the visual says something about the data is it an infographic.

Making your infographic work for you
To create a compelling infographic you need to think about all elements of the visual. Think about font size and format. To show context and add meaning, the size of font can convey effect size and/or frequency – bigger or heavier font shows 'more' or 'stronger', for example.

Think about typeface too – it can make a big difference. If I want to convey the message that a persistent cough over three or more weeks can indicate early-stage lung cancer, I can choose a 'jagged' font to 'show' the cough and convey a health promotion message.

Emotion and metaphor are where infographics come into their own and allow us to credibly engage readers, deepening understanding and supporting decision-making. If eighty people will die each month due to a certain condition, an infographic allows you to represent that figure as a double-decker busload of people. Now, rather than a 'flat' number, readers 'see' the impact of the condition in the vernacular of their daily commute.

Similarly, if it takes an hour of exercise to burn off the calories in a sugary snack, an infographic can present this as the average flight time from London to Paris. Now your data is compelling and can support a healthier lifestyle choice.

These are simple examples but they show the thought process of 'getting at' an effective infographic. Once the thinking is right, the design can follow. Only by understanding both your data and your target audience can you be effective via your infographic.

When more isn't better...
Crowding the visual by showing every data point can irreparably reduce the efficacy of the piece. When deciding what you want to convey, the KISS rule holds (keep it simple, stupid!).

Readers only rarely engage with more than three key messages at a time. Focus on the one story your data tells and, if it's part of a larger finding, use the infographic as a taster... leave the reader wanting more and drive them to the full research paper or website for more in-depth information.

Where to next?
Some great places to start to get a feel for infographics and 'build your own' include: Infogr.am (http://infogr.am/), Piktochart (http://piktochart.com/) and Visual.ly (http://visual.ly/).

Remember, your infographic should be led by your data and connect with your reader in a way that increases understanding of the research context. When you achieve this it's worth a thousand words and then some.

Caroline Collins

Complementary ideas include 1, 4, 26, 34 and 40.

47 Twitter – your research in 140 characters

In communicating your research you are in effect telling a story, and as a storyteller it is imperative that you engage with your audience effectively, and as often as possible. The nature of research means that it is geared up for presenting your work in the form of peer-reviewed journals and at specialist conferences, meaning that your findings often gain exposure infrequently and to a very small percentage of people. You have no doubt worked tirelessly to produce the results that you are now proudly willing to share with the outside world, but how can you ensure that your audience is not simply a couple of rows in an otherwise empty auditorium?

Finding a way to communicate your research in a frequent manner, and to a relatively large audience, can be tricky, but thankfully there exists a widely used and freely available tool to assist you: Twitter. By developing an online presence through Twitter, it is possible to gradually build up a diverse set of followers who all have an interest in you and your research.

Working out exactly what to tweet can be difficult, but it doesn't have to be vague summaries of your thesis chapters; oftentimes a link to a relevant research paper or a memorable quote and how it links to what you are doing can be far more effective. And when you do have an upcoming paper or presentation, be sure to advertise it with a summarising tweet!

When presenting your research in this manner, begin by splitting your story into three sentences which answer the following questions: 1) why did you conduct this research? 2) what did you find? and 3) what does this mean? Read these sentences out loud a couple of times and see how they sound; do they adequately convey your research? If not then try again, repeating this process until you are happy. After you have three sentences that you are settled on try condensing them down into a single tweet ('txt-speak' is allowed), again reading it out loud to get a feel for it. Finally, once you have your 140 characters tweet them to your followers and see what they think, using this real-time feedback on your research to improve your own understanding of the story that you are telling.

Paraphrasing the great physicist Richard Feynman, and turning his sentence into a tweet, if all of scientific knowledge had to be summed up it would be as such: 'All things are made of atoms, small particles that move around in constant motion, attracting when they are apart, but repelling if together.' This one tweet demonstrates not only the depth of communication that is possible in 140 characters, but also the limitless potential of the debate that it could inspire.

Samuel Illingworth

Complementary ideas include 4, 5, 12, 19, 33, 36, 49 and 50.

48 Guest blogging

Are you an expert in your area of research, have a flare for communicating science and feel at home in the blogosphere? Here are some tips on how to become a guest blogger.

Best case scenario is you strike it lucky and are invited to guest blog for someone else. This is not all that common (it very much depends on the area) but it does happen so take up the opportunity if it comes your way.

Another option that does not require a lot of hard work is creating 'threads' on existing blogs. Some owners allow readers to do that with the caveat that these threads will be moderated before they are posted. These days, publishers and media owners understand the value and power of user-generated content so, provided you are able to deliver interesting and relevant material, you are very likely to make the grade.

If the above options do not work for you, then you may need to try harder and persuade others that you are the perfect guest blogger. How do you go about it? The steps below may take some time but the likelihood is that you are already doing some of them so don't panic. In any case, you'll have plenty of fun along the way and results are guaranteed. Here it goes:

Identify the people you want to become a guest blogger for. Ideally, you already follow them, understand what they are all about and are passionate about the same issues.

Use social media to promote their work. There is no better way of making yourself known to the blog owner than by publicly showing appreciation by means of, for example, a series of supportive tweets or retweets.

Occasionally ask a question or make a comment on any of the threads within your area of expertise. If you have your own blog and social media accounts, use the same name or pseudonym and, if technology allows it, put a link to your blog page. You can also select a story that received some attention on his/her blog and refer to it in your own blog. Let them know you are doing this and invite him/her to comment. If you haven't got a blog, link your comment to an online personal profile (e.g. LinkedIn) that clearly identifies you and your area of expertise.

Identify areas where you think you can contribute. For example, are there any gaps in that person's knowledge? Do you know useful people they don't? Are you attending events they haven't got time or resources for?

Then ask the question. Here's how you can start the conversation: 'I love your blog and have been following it for a long time. I'd very much like to make a contribution on topic X with a guest blog post as this is very much my area of expertise. My view is that...'

And congratulations, you're on! The best thing about this approach is that it will pay off regardless of whether you are limited on resources and simply want to contribute with the occasional guest blog or if you are willing to invest time and effort and your goal is to become a regular contributor and get paid for it.

Bibiana Campos-Seijo

Complementary ideas include 12, 28, 33, 35, 38, 47 and 50.

49 Crowdsourcing

Crowdsourcing, or citizen science, is both a form of public engagement and a research method. It is distributed problem solving, in which researchers broadcast an open invitation to the general public to participate in completing tasks which contribute to a research project in some small but meaningful way. An element of crowdsourcing can be introduced at any stage of a research project, from helping formulate the research question to gathering or contributing data, and analysing and interpreting it. The volunteer audience or 'crowd' are active collaborators, not passive subjects or consumers of research.

Drawing on the extensive resources and wisdom of 'the crowd' through collaboration between researchers and volunteers benefits both parties. Crowdsourcing offers a way to actively develop an audience's understanding of the research process, as well as the research findings, and to foster their sense of empowerment, engagement and ownership of research. A well-designed crowdsourced project can allow a lay audience firsthand insight into how research is done and can also enable a project to draw on fresh perspectives, encounter serendipitous discoveries or handle issues far larger than a limited team of researchers could ever tackle. Crowdsourcing has been used in a wide range of disciplines, from biological and physical sciences to arts and humanities. For diverse examples of crowdsourced research projects you can take part in, from transcribing manuscripts to analysing bat calls and identifying galaxies, see the Zooniverse website, a portal which hosts such initiatives: https://www.zooniverse.org/

A crowdsourced project relies on excellent communication of the research subject, sound ethics, inspiring aims and methods, and clarity to ensure that the volunteers know what they are doing and can make methodologically robust contributions. For some research, crowdsourcing is a major part of the project design, used to handle large data sets and reaching a potentially vast audience through impressive interactive online interfaces such as Zooniverse. Crowdsourcing on this scale needs to be built into a research proposal, to ensure funding and resourcing for a sophisticated online infrastructure, and you should seek advice from your computing service.

Crowdsourcing can be used in smaller scale projects, however, with smaller 'crowds' and less expensive and technologically complex interfaces, such as social media. In the late nineteenth century, entries for The Oxford English Dictionary were crowdsourced by post! You might for example ask a 'crowd' to help you formulate your research question by Twitter, report experiences or encounters with your research interest on a simple Google form or map, or help refine a survey design using a wiki.

Consider how large and diverse the crowd needs to be, and cater for varying levels of involvement from your volunteers. Plan your publicity and networking strategy to ensure you're reaching beyond your immediate contacts. Contributors should be able to participate with their general knowledge or knowledge that can be learned in a very short time from the project, and with no specialist equipment. The task itself should be small but meaningful, discrete and quickly completed. Participants should never feel exploited or misled, but should gain some sense of reward. Reward could be formal acknowledgement, satisfaction in seeing the final results, a greater understanding about how research is conducted, learning about the subject, or maybe the experience of being part of a community of researchers and other volunteers, perhaps strengthened by a project forum or blog.

Helen Webster

Complementary ideas include 5, 12, 19, 24, 25 and 47.

50 Digital curation – collecting and sharing online resources

As twenty-first century researchers, we often create digital teaching tools or find online information that we would like to share with students or colleagues. We might want to enhance a reading list, cover a particular topic in depth or have a website to display our work. We can do this by blogging, but it can be labour intensive to write posts regularly. Another possibility is *digital curation*, a way of gathering, collating and presenting online resources such as web pages, social-media postings, images and videos. Such digital collections are easy to create and easy to find via search engines, allowing us easily to engage readers and potentially gain an audience outside of academia for our research and interests. We often already make collections of *physical* resources such as articles, photographs or slides; an online digital collection, chosen and vetted by us, is a logical extension of these activities.

Types of digital curation tools

Digital curation tools fall broadly into three types: content curation (e.g., Scoop.it or Kippt), bookmark curation (e.g., Diigo or Delicious) or social-media curation (e.g., Storify or Paper.li). One tool will likely be better than another, depending on your needs and what you wish to curate. For example, if you wish to bring together interesting websites on a certain topic in a professional magazine-like format, then *content-curation* tools would be good. If you are reporting and commenting on a particular event, such as conference, then a *social-media* curation tool might be better. If you'd just like to provide a list of links for further exploration to your students, then *bookmarking* tools are excellent. Experimentation will help you find the tools that are best.

How to set up digital curation tools

Despite their differences, digital curation tools often work similarly. You usually need to register and create a user name and password. Many sites give you the option to login via Facebook or Twitter. Making an account is required to be able to save your content over time. Many online curation tools also come with a bookmarklet for your browser's bookmark bar, allowing you easily to snip content and add it to your portfolio. Many curation tools also remember preferences and topics of interest and offer suggestions for further content to collect. If you have a blog or other website, it's often possible to set up a feed or a widget to display your curated content directly on your site.

A few words of caution
The ease of snipping content with these tools can separate content from its sources, making it difficult to determine provenance. You should make an effort to attribute sources by providing a link back to original content whenever possible. Also, poor search terms and other metadata may mean that automatically suggested content is not always appropriate. You need to experiment with the best terms and searches to yield relevant information. Finally, some effort must be taken to make sure that content is regularly kept up to date, as your audience will hopefully return to the site for the latest research and news in your topic area.

Digital creation tools easily create a professional-looking presence on the web. A carefully curated site with quality content can be a unique place where you can present your research or scholarly activities, engage in a dialogue with your students and colleagues, and be a long-term repository for your own important web content.

Meg Westbury

Complementary ideas include 12, 14, 16, 19, 33 and 35.

51 Demonstrating how your research makes you employable

You are a professional researcher, with an excellent track record, a raft of experience and some very impressive qualifications. When it's time to find yourself a new role, however, the key question becomes how to put an application on paper that allows you to stand out from all of the other professional researchers with their imposing qualifications and spectacular experience.

You need to acquire a set of evidenced data to prove that you have the knowledge, skills and attributes (KSA) that make you an attractive proposition. Simple as it may sound, just being able to elicit, explain and evidence (3E) your skill-set will be enough to make you 'better than most' researchers in this crowded job market.

Make yourself a 'bucket-list' table, with the columns being your KSA. Include:

1) Knowledge – the facts that you know, e.g. how to use 'Endnote', access European archives or submit successful ethics committee approvals.

2) Skills – your expertise. All good researchers will be effective project managers and impactful communicators, so how can you evidence your skill-set? Did you plan, deliver and graduate with your PhD within 36 months? Did you win an award for outreach? It can help to think of additional skills in terms of key areas, such as the standard (and advanced) research skills and techniques of your specialism, research management skills, personal efficacy, communication skills and networking skills. Do your skills research – the internet is full of examples of skills lists and definitions, and finding evidence from your career to date is simply a mapping and data collection exercise.

3) Attributes – your personal characteristics. You are looking for those that will make you both easy to employ and a valuable

colleague. Great examples include professionalism, tenacity, punctuality, etc., but the list is long. This is not the time to be modest; you need an honest assessment of what makes you good at what you do.

Once you have your history 'defined' in your columns, we add 'value' in the rows of your table, using the '3E' technique:

1) Elicit all that you have to offer – descriptors of your experience, taking in all that you have studied (at modular level, don't forget professional courses), all of the posts that you have held, the projects that you have worked on and the achievements that you are proudest of.
2) Explain what you have been responsible for and what you have delivered (your research outputs and the impact you have made).
3) Evidence – think in terms of numbers and data wherever possible, for example citations, publications, lectures and seminars delivered, percentage improvements as a result of your work, and so on.

Once you have your table all you need to do is to work out which of the KSA will be most applicable to the specific job that you are applying for, and then make sure that you elicit, explain and evidence their relevance to your future employer.

Caron King

Complementary ideas include 3, 22, 30, 37 and 41.

52 Body language

You're nervous. You're sitting down, your shoulders are hunched and your heart is racing. The time to present your research is drawing close.

Many pieces of research are masked by the diffident manner of the presenter. Presenting oneself physically as confident can improve the communication of research.

There are many techniques for doing this but the following tips will take you a long way: be centred; be big; and above all, be connected.

First, be centred
Picture a spirit level: a tube filled with liquid and a single bubble that sits between two bold black lines when the tube is held exactly level. Any movement out of equilibrium and the bubble will shift out of place. Now imagine that you have one of these at your centre of gravity, right in your core, between your navel and your spine. In order to keep the bubble in the centre, you need to keep your core stable.

Place your feet hip-distance apart, your weight evenly distributed between your feet. Focus for a moment to ensure that bubble is stable, not tipping this way or that, forward or back. Breathe steadily, and if you start to feel anxious during your presentation, refocus on the bubble's stability.

This will be familiar if you have done any yoga. This stable pose projects confidence as well as helping to keep you calm inside.

Second, don't hold yourself small
Generally the human reaction to fear is to retreat into oneself, physically shrinking by drawing tense limbs close to the torso, hunching the shoulders. An effective way to project confidence is to avoid this by doing the opposite: make yourself big!

Before you present your research, go somewhere private and make yourself as big as possible. Stretch your arms upwards, as if you are cheering. Hold that pose. Smile. You are opening your airways (oxygen is always a plus) and giving yourself a chance to feel celebratory – the opposite of scared.

When you drop the pose and leave your private spot, hold on to that feeling. Keep your shoulders relaxed and back and hold your head high. Breathe. Now open your mouth and speak.

Third, and most importantly, connect
The danger is that your centred, big posture will come across as arrogant or ridiculous, thus irritating or disconcerting the audience. Avoid this by remembering that the audience is not an amorphous blob of humanity. It comprises a number of individuals. Consider what these individuals need and connect with them. That's you, connecting with her. And him. And him. And her. What is natural for you to do when communicating? Do that.

It is important to discover for yourself when to break the 'rules' for the sake of connecting naturally with the people in the room. Communicating research with confidence is not about where you focus or how straight you stand. It is about creating a connection that allows ideas to flow from you to whomever you wish to communicate with.

Helen Lawrence

Complementary ideas include 10, 14, 16, 27, 32 and 39.

53 Mastering Q&A sessions

A successful question and answer session relies upon your ability to recognise the motivation behind questions, so that you can respond accordingly while ensuring that the wider audience remains supportive. However diverse the questions, there are really only five types of questioner.

Confused

It can be embarrassing when someone has clearly misunderstood a point that you made or, worse still, has completely failed to hear something you said. It is your job to avoid this embarrassment. NEVER say 'As I think I just mentioned' or 'As I said earlier'. Simply answer the question as if it were perfectly acceptable. The audience will love you for such a neat solution to a potential problem and the questioner will not have the horror of realising the mistake until after the event.

Oratory

Some questions are actually mini-lectures in disguise. Any question which starts with something like 'What I found particularly interesting was...' is likely to segue rapidly into a lengthy exposition of the questioner's view of your topic. Be patient if you can: an interruption risks further oration. At the end there will be a minor question: answer it briefly and move on to the next questioner determinedly. You need not mourn the loss of time: there will be a chance to chat to others interested in your research later in the day.

Aggressive

Academia, as in every walk of life, contains unpleasant people. An offbeat question that is purely designed to trip you up and to which you do not know the answer – usually to make the questioner feel good – is not worth engaging with too deeply, as this is not a fight that you can, or would want to, win in a head-to-head confrontation. Instead, thank the questioner for the question, explain that you need to look into it more fully and offer to email that person the next day. Smile winningly, then break eye contact.

Unexpected

Friendly questions you were not expecting are rare, but if you get them it is always worth asking the questioner what he or she thinks. It is perfectly acceptable to make a question into a more general discussion for a few minutes.

What can also be unexpected is the silence that sometime arises, when nobody seems to have a question. This is particularly awkward if you have an aggressive questioner in a small group from whose persistent questioning you would like to escape. It always helps to have a friendly colleague on hand when you present, someone ready to step in and ask a helpful, supportive question if silence prevails for too long.

Helpful

Always a delight. Make the most of easy-to-answer questions, but try not to let this become a mini-lecture of your own: you have already given the presentation.

To keep the audience with you throughout the Q&A session, always begin an answer facing the questioner, make sure that you move your eye contact to include the whole audience after the first sentence, then return to your questioner at the end and ask (in words or with a questioning smile) if you have answered the question adequately. Only allow one supplementary question. Regardless of the tone or content of any question you receive, resolutely assume for these few minutes that everyone in the room is friendly, supportive and impressed.

Lucinda Becker

Complementary ideas include 1, 10, 18, 31 and 39.

Notes on contributors

Ruwayshid Alruwaili is a PhD student in the Department of Language and Linguistic Science at the University of York. He works primarily on second language acquisition.

Dr Catherine Armstrong is Lecturer in Modern History, Loughborough University. Catherine is a colonial American historian with expertise in the connections between Britain and the south-east of North America; also a book historian focusing on the connection between print culture and identity.

Esther Barrett is a PhD candidate and eLearning Advisor at Jisc RSC Wales, where she supports the innovative use of technology in teaching and learning in post-16 education in Wales.

Dr Lucinda Becker is Associate Professor in the English Literature Department of the University of Reading. She teaches undergrads in the English Literature department and postgraduates in several other departments. Lucinda is also a professional trainer, primarily for engineers, scientists and lawyers.

Dr Lucy Blake is a Research Associate at the Centre for Family Research, University of Cambridge. Lucy researches family relationships and psychological well-being in different kinds of families. She is also responsible for making podcasts about the Centre's research.

Dr Nancy Bocken is Lead Researcher in sustainable business development at the University of Cambridge. Nancy conducts multidisciplinary sustainability research, which has an impact on the way businesses operate.

Debbie Evonne Braybrook is a PhD student at the Centre for Men's Health at Leeds Metropolitan University. She has experience as a research assistant in the third sector and in academia, and is currently completing her PhD.

Nicola Buckley is Head of Public Engagement at the University of Cambridge. In this capacity she manages the team organising the Cambridge Science Festival, the Cambridge Festival of Ideas and other public engagement events and activities across the University.

Dr Victoria Burns is Senior Lecturer in the School of Sport, Exercise and Rehabilitation Sciences at the University of Birmingham. Victoria completed a British Science Association Media Fellowship in which she worked as a science journalist at the *Irish Times*.

Dr Bibiana Campos Seijo is the Editor of *Chemistry World* and Magazines Publisher at the Royal Society of Chemistry. She previously completed a PhD in chemistry and ran her own e-learning business.

Dr Eleanor Carter is Clinical Research Fellow in the Division of Anaesthesia at the University of Cambridge. Eleanor is a practising anaesthetist with academic roles in brain injury research and medical education.

Dr Caroline Collins is an account manager and public affairs and PR consultant. As an author, public affairs and communications specialist Caroline has successfully delivered national and European health, pharmaceutical and lobbying campaigns over the last eleven years.

Dr Irenee Daly is a psychology lecturer at DeMontfort University where she is also a member of the Reproduction Research Group.

Brennan Decker is a Cambridge Scholar at the National Institutes of Health. As an MD/PhD student researcher, Brennan studies genetic factors that influence an individual's risk of developing cancer.

Dr Emma Gillaspy (www.emmagillaspy.com) is a researcher development consultant. Her current portfolio includes Vitae North West Hub Manager and Faculty PGR Development Lead in the University of Manchester. Emma uses her training, coaching and change management skills to support researchers and academics with their professional and career development.

Dr Carina Girvan is a lecturer at Cardiff University. Carina is interested in transformative technology-enhanced learning experiences, most recently exploring constructionism in non-goal orientated virtual worlds such as Second Life.

Dr Aoife Brophy Haney is a Senior Researcher at ETH University in Zürich. Aoife completed her PhD at the Cambridge Judge Business School. She researches how businesses innovate in response to sustainability challenges.

Erika Hawkes is Postgraduate Researcher Development Officer at the University of Birmingham. Erika researches, designs and delivers development activities for 3000-plus postgraduate researchers, from presentation skills to speed reading via writing summer schools and research exhibitions.

Anthony Haynes is Creative Director at The Professional and Higher Partnership (www.professionalandhigher.com). Anthony mentors researchers on how to communicate their research in order to get published, win grants, get jobs, and engage wider audiences.

Dr Steve Hutchinson is an independent consultant who works internationally, specialising in leadership and personal effectiveness training and development (www.hutchinsontraining.com).

Dr Samuel Illingworth is a lecturer in Science Communication at Manchester Metropolitan University, where he uses his expertise in public engagement and theatrical technique to develop the effective communication skills of scientific researchers.

Dr Steve Joy is a careers adviser at the University of Cambridge. Steve offers careers education and guidance to researchers in the arts, humanities, and social sciences.

Hubertus Juergenliemk is a final year PhD student at the University of Cambridge, focusing on the EU's security and development policy in Kosovo and DR Congo. Hubertus works as Adjunct Professor at Vesalius College Brussels.

Dr Caron King is Operations Director at the Mindset Method (www.mindsetmethod.com). Caron is a change and project manager and facilitator who focuses on enabling individuals and teams to be as effective at work as possible.

Jim Krane is a Fellow for Energy Studies at Rice University. As a journalist he has contributed to the *Wall Street Journal, Financial Times,* and The Economist Intelligence Unit.

Dr Eva Lantsoght is an Assistant Professor at Universidad San Francisco de Quito USFQ where she teaches structural engineering courses. She is also a part-time researcher at Delft University of Technology in the field of concrete bridges.

Dr Helen Lawrence is an independent consultant (http://www. helenlawrencetraining.com/)who works throughout the UK and beyond, seeking to help researchers and others become more effective.

Dr Amelia Markey is Associate Medical Writer, QXV Communications. Amelia is a recent analytical science PhD graduate from the University of Manchester who has moved into a career in medical communications as a medical writer.

Dr Agata Mrva-Montoya is Publication Coordinator at Sydney University Press. Agata has worked at Sydney University Press since 2008 in a role combining editing, project management, ebooks and social media.

Hannah Perrin is a PhD candidate and Assistant Lecturer at the University of Kent. Hannah is a researcher in the sociology of veterinary medicine at the School of Social Policy, Sociology and Social Research, with interests in professional identity and occupational socialisation.

Han Pham is Future Cities Experience Strategist at Intel Labs Europe. Han helps envision the future of how we will live, work and play in our future cities by designing engaging and sustainable urban experiences for Intel Labs Europe through the lens of behaviour, culture and design.

Dr Dan Ridley-Ellis is a Principal Research Fellow at Edinburgh Napier University. Dan is one of the UK's experts on wood properties and timber grading, and he is a central organiser of Bright Club Edinburgh and Bright Club Scotland.

Dr Sara Shinton is Managing Director at Shinton Consulting Ltd (www.shintonconsulting.com). Sara's consultancy improves the career awareness and employability of academic researchers.

Karen A. Souza is currently a doctoral student in psychology at City University London and a Visiting Scholar at the Institute of Criminology, University of Cambridge.

Joel Suss works with the LSE Public Policy Group and is Managing Editor of the British Politics and Policy blog.

Dr Helen Webster is an academic developer at Anglia Ruskin University. Helen works with academic teaching staff to develop good practice in learning, teaching and supporting students in higher education.

Meg Westbury is the Librarian for Wolfson College, University of Cambridge. Meg is an information specialist who closely follows trends in usability, marketing, and new and social media.

Monica Wirz is a management consultant at ClueTrain Consultancy and a partner at Effective Boards. After twenty-six years of international experience she has returned to academia where she is about to defend her doctoral thesis at the University of Cambridge.

Rebecca Woods is a PhD student at the Department of Language and Linguistic Science, University of York. Rebecca is interested in research communication following a stint as a librarian in the Faculty of Classics, University of Cambridge.